Research Methods in Dentistry

Fahimeh Tabatabaei • Lobat Tayebi

Research Methods
in Dentistry

 Springer

Fahimeh Tabatabaei
School of Dentistry
Marquette University
Milwaukee, WI, USA

Lobat Tayebi
School of Dentistry
Marquette University
Milwaukee, WI, USA

ISBN 978-3-030-98030-6   ISBN 978-3-030-98028-3   (eBook)
https://doi.org/10.1007/978-3-030-98028-3

This Springer imprint is published by the registered company Springer Nature Switzerland AG
The registered company address is: Gewerbestrasse 11, 6330 Cham, Switzerland

Preface

Advances in dentistry are the result of great dentists and researchers' collaboration efforts in the past, present, and future. Without their efforts, dentistry would not have achieved its present status. New science and technologies such as smart biomaterials, biomimetic, functionally graded materials, tissue engineering, and 3D printing are rapidly finding their way into dentistry. However, many questions remain about the treatment of specific oral defects such as critical-sized ones or etiology of oral diseases and the most effective ways to prevent or treat them. Any dental material (new resin composite), pharmaceutical (triple antibiotic), and preventive product (varnish), or new treatment protocol (guided bone regeneration) needs to be validated by dental researchers. Having an effective role in the investigation of unresolved challenges and participation in the evolution of dentistry require employing a reasonable research methodology. Therefore, learning the methodology of research is an integral part of dental education to prepare dentists for a dynamic future that is dependent on generating new knowledge in this field.

Teaching the research methodology in dental schools is not only for conducting research but also for obtaining the critical thinking and reading skills necessary to incorporate new evidence into the clinic. After years of teaching and mentoring dental students, the authors of this book realized that the growing volume of research articles during the recent years has overwhelmed and confused many students in their research activities. They usually ask questions about how to read and cover all these papers, how to find the points needed in research among hundreds of existing documents, or how to be sure that what they are doing in their project has not already been done by other investigators. This book aims to help dental students and future dental practitioners to overcome such challenges and be competitive in today's fast-growing research environment.

The book presents the research methodology in dentistry in seven chapters. The chapters complement each other but can also be used independently. The first and second chapters explain the key concepts and common approaches in dental research, both in basic science and clinical dentistry. The third chapter familiarizes the readers with evidence-based research in dentistry and how to write a systematic review. The fourth chapter explains the process of designing and presenting a proposal. Chapters 5 and 6 discuss reporting the results of scientific studies and managing the references. The final chapter is about ethics in research, highlighting the significance of adherence to ethics in animal and human studies. This book provides practical guidelines for dental researchers; however, it can also be beneficial for researchers in other fields (e.g., medical sciences and biomedical engineering).

Milwaukee, WI, USA
Fahimeh Tabatabaei
Lobat Tayebi

List of Abbreviations

3D	Three-Dimensional
ADA	American Dental Association
AFM	Atomic Force Microscopy
ANOVA	*Analysis of Variance*
APA	American Psychological Association
ARRIVE	Animal Research Reporting of In Vivo Experiments
CAD/CAM	Computer-Aided *Design*/Computer-Aided Manufacturing
CI	Confidence Intervals
COI	Conflict of Interest
CONSORT	Consolidated Standards of Reporting Trials
COPE	Committee on Publication Ethics
COVID	Coronavirus Disease
DOAJ	Directory of Open Access Journal
DOI	Digital Object Identifier
EARR	External Apical Root Resorption
EBD	Evidence-Based Dentistry
EDS	Energy Dispersive Spectroscopy
EMBASE	Excerpta Medica dataBASE
Er: YAG	Erbium-doped Yttrium Aluminum Garnet laser
FDA	Food and Drug Administration
GBR	Guided Bone Regeneration
GIC	Glass Ionomer Cement
GPa	Gigapascals
GTR	Guided Tissue Regeneration
H_0	Null Hypothesis
HIPAA	Health Insurance Portability and Accountability Act of 1996
IADR	International Association for Dental Research
ICMJE	International Committee of Medical Journal Editors
ICTRP	International Clinical Trials Registry Platform
IEEE	Institute of Electrical and Electronic Engineers
IgM	Immunoglobulin M

IL	Interleukin
IPD	*Implant* Probing Depth
IRB	Institutional Review Board
ISO	International Organization for Standardization
JDE	Journal of Dental Education
JDR	Journal of Dental Research
JDS	Journal of Dental Sciences
JPEG	Joint Photographic Experts Group
KAP	Knowledge, Attitude, Practice
MEDLINE	Medical Literature Analysis and Retrieval System Online
MeSH	Medical Subject Headings
MPC	2-Methacryloyloxyethyl Phosphorylcholine polymer
MTA	Mineral Trioxide Aggregate
MTT	3-(4,5-dimethylthiazol-2-yl)-2,5-diphenyl-2H-tetrazolium bromide assay
NCBI	National Center for Biotechnology Information
Nd: YAG	Neodymium-doped Yttrium Aluminum Garnet laser
NIDCR	National Institute of Dental and Craniofacial Research
NIH	National Institutes of Health
NiTi	Nickel-Titanium
NOS	Newcastle-Ottawa Scale
NSF	National Science Foundation
OIERR	Orthodontic-Induced External Root Resorption
PDF	Portable Document Format
PDL	Periodontal Ligament
PICO	Patient/Population, Intervention, Comparison, Outcomes
PMA	Premarket Approval
PMID	PubMed identifier or PubMed unique identifier
P–P	Percentile-Percentile
PRISMA	Preferred Reporting Items for Systematic Reviews and Meta-Analyses
qPCR	Quantitative Polymerase Chain Reaction
Q–Q	Quantile-Quantile
RCT	Randomized Controlled Trial
RevMan	Review Manager
RF	Risk Factor
RIS	Research Information Systems
SD	Standard Deviation
SE	Standard Error
SEM	Scanning Electron Microscopy
SJR	SCImago Journal Rank
SPSS	Statistical Package for the Social Sciences
STARD	Standards for Reporting of Diagnostic Accuracy
STAs	Sequential Thermoplastic Aligners
STROBE	Strengthening the Reporting of Observational Studies in Epidemiology

SYRCLE	Systematic Review Centre for Laboratory Animal Experimentation
TIFF	Tag Image File Format
WHO	World Health Organization
WMA	World Medical Association
WWH	What-Who-How
XML	Extensible Markup Language

Contents

Chapter 1
Introduction to Dental Research

1.1 What Is Research?

Scientific research is a rational approach that allows the examination of problems to be solved and the discovery or formulation of precise answers to *questions*. It is built on the work of other scientists, can be generalized, and generates new questions. The purpose of the research could be to earn doctorates or masters, meet curiosity, make discoveries, provide innovative solutions to complex problems, develop new products, save costs, advance science, acquire new knowledge, or improve the well-being of society. The ultimate goal of the research is to create the knowledge essential for improving health. Without this knowledge, this action is impossible because it has no logical basis.

The process of research is characterized by the fact that it answers questions in an *organized*, *scientific*, and *systematic* manner.

- Questions are required for research, and research always arises from a problem to be solved. Without a specific question, your research will be aimless. In other words, there is no research where there is no question. Therefore, by stating your question, you will clarify what your study intends to address. Finding the answer to the research question will be the objective of your research.
- *Organized* research begins with a plan or proposal. Non-planned research is often inefficient, and it may never lead to the researcher's correct answer.
- The *scientific* method means that the study is based on reproducibility, feasibility, and reliability, which we will explain further in the following chapters.
- Finally, using a *systematic* process means following clearly defined steps in the research to achieve an accurate answer: If your approach is *deductive*, you should: (1) formulate the research question, (2) predict the answer to the research question in the form of a hypothesis, (3) design a plan, (4) test the hypothesis, and (5) evaluate and report the results which confirm or reject your hypothesis. For example, after treating a carious-exposed tooth using calcium hydroxide and composite, you noticed the dentin bridge formation under restoration. Your ques-

© Springer Nature Switzerland AG 2022
F. Tabatabaei, L. Tayebi, *Research Methods in Dentistry*,
https://doi.org/10.1007/978-3-030-98028-3_1

tion is: What is the reason for the formation of the dentinal bridge? You hypothesize that calcium hydroxide caused it. To test this hypothesis, you design a plan. After examining the effect of calcium hydroxide on dental pulp stem cells, you can report that your hypothesis has been confirmed or rejected. In an *inductive* approach, you should: (1) carefully observe a situation or an individual case, (2) gather the findings, (3) authenticate the underlying characteristics, (4) find a pattern in other cases, and (5) formulate your theory. The theory of gravity is an example of an inductive approach. By observing the fall of an apple from a tree and the recurrence of this fact in other objects, Newton concluded that there must be a gravitational force that causes this pattern to happen.

Therefore, one of the basic steps in any research project is to choose the *research approach*. This choice depends above all on the nature of the problem and the phenomenon being studied. As the deductive approach is more prevalent in dental research, we will focus on this approach in this book. Similarly, researchers should be familiar with the different types *of research* and know their application in their field of study.

Before dealing with the types of research, it is necessary to understand the exact meaning of the key concepts in the research literature, which are also used extensively in this book.

Method: The word *method* may refer to a tool, instrument, experiment, technique, and test to set up, conduct, collect, analyze, and evaluate data for answering the research question. Observations, experiments, and statistical approaches are all examples of the methods employed to conduct the research. For example, you may work with a "universal testing machine" to obtain the compressive strength of your new composite samples; you can use an MTT assay (3-(4,5-dimethylthiazol-2-yl)-2,5-diphenyl-2H-tetrazolium bromide assay) for evaluating the cytotoxicity of these composite samples; or you may employ "one-way *analysis of variance* (ANOVA)" for analyzing your data.

Methodology: The research methodology is broader than the method and refers to several methods used to efficiently investigate the research hypothesis and achieve the objective. Accordingly, the methodology may include subsections like "Research Design," "Protection of Human Rights," "Instrumentation," "Data Collection and Analysis." Methodology implies the correct application of the method, explanations behind the decision of your specific methods (the rationale), and the recognition of the criteria for choosing a method. Methodology asks: Why did you use the compression test? Is it the best way to solve the problem? If your material is brittle, you should consider the diametral compression test. Therefore, "methodology" or "approach" is the investigation's work plan and justifies and supports a particular method's choice by citing relevant sources.

Using a hook is a *method* of fishing, while how to use it is your *methodology*. The methods used should be valid and reliable, which means that they must precisely measure the expected properties and, if repeated, lead to the same results. Suppose that you are studying the effect of a substance on craniofacial regeneration. In that case, the method you are using should be such that anyone who uses it achieves the same results. Sometimes, the method is valid and reliable, and you are performing it correctly, but it does not answer your question, which means that your methodology is incorrect. When your objective is to investigate the effect of mineral trioxide aggregate (MTA) on the osteogenic differentiation of stem cells, the MTT test will not answer your question even if it is done correctly. The MTT test is not a proper method for measuring differentiation. For this purpose, you need to assess the osteogenic marker expression by real-time quantitative polymerase chain reaction (qPCR) [1]. Therefore, your research design guarantees that the correct choice of methods is genuinely and solely responsible for the results and leads to the correct answer to the research question. To employ a proper method, you need to understand the research problem accurately. Then, the selected methods should be analyzed rigorously (clarifying the logic behind them) to ensure the results' validity, reliability, and credibility.

Variable is the measurable factor that can vary during research, can have more than one value, and is involved in answering the research question. The variable must also be directly observable (sex, color of eyes), or if it is not directly observable (time, temperature, proportion, pH), it can be measured by another measuring instrument (pH meter). The cause, risk factor, or the variable that can be manipulated in the research method is called the *independent variable*. The *dependent variable* is the effect, outcome, or the variable that responds to changes and would be analyzed to meet the objective. The independent variable can be changed/manipulated/controlled at given levels/doses/values/types by the researcher. Therefore, you can purposefully change the independent variable, which causes a change in the dependent variable (the reverse is impossible, which means the dependent variable cannot cause any change in the independent variable). The dependent variable should be measurable, and you should know how to measure it. When you study "the effect of MTA on the osteogenic differentiation of stem cells," MTA is the independent variable, and the dependent variable (osteogenic differentiation) is measurable.

You can observe or measure the dependent variable (effect) to report the correlation or association between the independent and dependent variables or find a difference between them. We say that an association (correlation) exists between two variables when a change of one variable coincides with the shift in the other variable. An association can be positive or negative and proportional or non-proportional. The association is causal when a change in the independent variable (exposure) leads to or causes a difference in the dependent variable (disease). For example, you may assess patients regarding the use of restorative materials and the location of caries. If the results show that restoration selection depends on tooth type and surfaces, there is an association between the two variables [2]. When you assign subjects to groups and compare them to find the difference, the independent variable is the group classification; the dependent variable is that on which they differ. For instance, you may distribute several extracted teeth to four groups restored with different bulk-fill resin-based composites (independent variable) and compare them in terms of internal adaptation (dependent variable) [3].

Hypothesis, in simple terms, is a statement in response to the research question and should be stated based on the researcher experiences and the findings of previous studies. This statement can be proven or rejected. To write a hypothesis, the researcher must determine the variable or groups and predict the relationship or the difference between them. The *null hypothesis* (H_0) usually predicts no difference between the control and the study group, while the *alternative hypothesis* assumes a difference between groups or association among the variables. Hypothesis plays a guiding role in the research process and is only validated after experimentally testing. The role of hypotheses is to show the researcher the general direction for conducting research; they prevent the study of resources not related to research, help in the correct determination of the methods, and provide a framework for interpreting the information gathered and drawing conclusions. Based on the specific aims of your research, you may have several hypotheses. In a good research design, rejecting a null hypothesis will construct another null hypothesis, and the study will continue.

The *participants/samples* are drawn from a *population* with determined characteristics. As the population of interest is usually substantial, and it is impossible to work directly on all of them, most research studies involve observing a *sample* from a defined population. Sampling is the process of choosing portions of the population to observe and study. The *sampling method* can be probable (random) or non-probable. It is essential to be aware of the difference between *random sampling* and *random assignment/allocation*. While *random sampling* is considering a known chance (greater than zero) of participation for every member of the population to be selected as subjects, *random assignment* means that after selecting subjects, you should also determine a method of randomly assigning the sample into the study/control groups.

Finally, one of the most crucial parts of the research is the *researcher*. The researcher is any person carrying out studies respecting a rigorous methodology to provide new knowledge. The researcher only deals with facts within a framework defined by the scientific community. A researcher needs to be passionate about his/her work, have a good plan, be patient while doing research, and be persistent in achieving the goal. Among the critical qualities of researcher leading to success in research, we can cite:

- A curious mind to find new facts.
- Integrity for the value of the scientific method.
- An analytical mind capable of practicing critical thinking.
- Receptivity to criticism at the professional level.
- Open-mindedness and the ability to see the meaning of unexpected observations.
- Objectivity.

Now that we are familiar with the terminology of research, we can more easily get acquainted with different types of research.

1.2   Types of Dental Research

The type of research significantly impacts the method of data collection and analysis, and if not chosen correctly, it can lead to erroneous results or an inability to answer the research question. Therefore, it is imperative to decide on the type of research before starting the project because you cannot change it during the study. Let's see what the primary types of research are.

1.2.1   Based on the Time

One of the criteria for research is the issue of time. Based on the time, the researcher observes the subjects/patients at a given time (cross-sectional studies) or during a continuous period (longitudinal studies) (Fig. 1.1).

In *cross-sectional* or prevalence studies, you want to define "what is happening." Like taking a picture with a camera, you need to observe all samples of different groups once at the same point in time or over a short period at the present moment. When you simultaneously compare the salivary cytokine levels of periodontally healthy subjects and subjects with chronic periodontitis, your study is cross-sectional. Here, you measure the dependent and independent variables (the exposure and the outcome) at the same point in time and cannot predict which came first (the increase of salivary cytokine or chronic periodontitis) [4]. Consequently, you cannot infer causality. Another application of prevalence studies is the estimation of the prevalence of the outcome of interest but not its occurrence. For instance, you can observe the presence of common dental anomalies in a sample of non-orthodontic growing subjects by examining their panoramic radiographs [5]. You can also use this type of study for knowledge and attitude evaluations by conducting surveys [6]. If data collection is performed several times in cross-sectional studies, it can be named a *pseudo-longitudinal* study. For instance, you select a sample of adolescents to evaluate the improvement in the toothbrushing frequency and its associated factors over three studies [7].

In *longitudinal* studies, the same group is observed continuously or repeatedly over a period (like a movie) to examine changes over time and understand the relationship between variables. This study is more accurate than cross-sectional and can be utilized for discovering the predictors of diseases. When you select two groups of subjects with/without a high level of a salivary cytokine and recall them

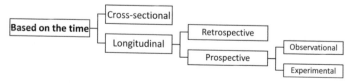

Fig. 1.1 Classification of research studies based on the time

periodically to evaluate the development of periodontitis in each group, subjects are enrolled in a longitudinal study. The longitudinal research can be retrospective or prospective:

- *Retrospective* studies concern any study for which exposure has occurred in the past, and data will be collected after the exposure or intervention through interviews or recorded information of patients. This method aims to analyze and interpret past events to predict similar events in the future. In other words, this method expresses "what happened." Assuming that you are selecting a group of patients who previously received direct posterior resin-based composite restorations, you are conducting a retrospective longitudinal study if you evaluate the survival rate of their restoration [8].
- In the *prospective* study, the research starts now, and the data will gradually be collected to determine the exposure. You aim to express "what will happen" regarding the etiology of a disease or the efficacy of an intervention. For instance, you can analyze salivary biomarkers of bone loss in two sample groups at risk or no risk of periodontal disease to see in which group more bone loss will occur [9]. Prospective studies could be *interventional (experimental)* or *non-interventional (observational)*. Non-interventional studies involve only questionnaires completed at a consultation or routine follow-up of participants. On the other hand, in interventional (experimental) studies, the intervention is the researcher's responsibility. If a group of patients recently received resin-based composite restorations, and you only examine their restorations every 6 months for 12 years, your study is observational, while, if you select a group of patients, repair their teeth by composite restorations after assigning them to different groups, and then follow up with them over a period, your study is prospective and interventional.

1.2.2   Based on the Location

Based on the location of the data collection, the research can be divided into two categories: field research and library research (Fig. 1.2).

Field research involves the researcher's own observations and data collection. For example, to evaluate "the salivary cytokine levels in subjects with chronic periodontitis and periodontally healthy individuals," you need to select the subjects, collect their saliva, measure the cytokine level in the samples, and then analyze the

Fig. 1.2 Classification of research studies based on the location

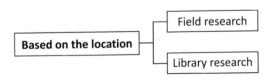

collected data. This type of research in which the researcher gathers data directly is also called a *preliminary* or *original study*.

In *library research*, the data is already available and should only be categorized. *Systematic review* and *meta-analysis* are the common forms of library research, where the researcher refers to the libraries or bibliographic databases and uses existing data that has already been submitted. Such studies that are based on the results of primary studies are also called *secondary studies*. It should be noted that the library research does not mean the obligation to be in the library, but rather the commitment to the library data. For instance, you can extract information from the histopathological data available in the Oral Pathology database [10].

1.2.3 Based on the Type of Data

Depending on the data type, two kinds of studies can be identified: *qualitative* and *quantitative* (Fig. 1.3).

If the information received or collected is not subject to statistical analysis but is based on an analysis beyond numbers, we refer to it as *qualitative* research. Here, you do not deal with numbers and statistics but rely on data explanation and interpretation of a small sample size. Images obtained from the optic, scanning electron, or fluorescent microscopes; spectra obtained from X-ray diffraction and Fourier transformation infrared spectroscopy; and a result obtained by cell and tissue staining are all examples of qualitative data.

In contrast to qualitative research, the emphasis in *quantitative* research is on statistics. It aims to collect quantifiable data and statistically analyze them to search for correlation (association) between variables or differences between groups. In this approach, sampling is critical to be representative of the population. Data should be analyzed using statistical methods, and significant or nonsignificant differences between groups should be statistically summarized and compared. Therefore, to ensure the meaningfulness and usefulness of quantitative data, it is essential to choose the analysis method before data collection.

1.2.4 Based on the Application

Based on the outcome or purpose of the study, the research can be categorized into three types: fundamental (basic), developmental, and applied (Fig. 1.4).

Fig. 1.3 Classification of research studies based on the type of data

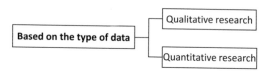

Fig. 1.4 Classification of
research studies based on
the application

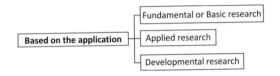

1.2.4.1 Fundamental or Basic Research

Fundamental research aims to identify new problems, discover facts, recognize phenomena, test theories, and generate knowledge. This research is concept-oriented and is the primary source to produce new knowledge for applied and developmental research. This type of research is about expanding the boundaries of knowledge without having concern about its immediate application. The keys to success in this type of research are new ideas, a highly motivated team, and access to modern technologies. The scientific community uses its results. This type of study is entirely theoretical and cannot be used immediately after the project accomplishment. It is more time-consuming and has strategic goals. Governments often support this research and encourage institutions to address these fundamental issues.

Fundamental research is usually conducted in a controlled environment such as a laboratory to strictly observe the samples and prevent the interaction of different variables.

> Examples of basic research in dentistry include:
>
> 1. How does *Streptococcus mutans* cause caries?
> 2. How do medications induce dry mouth (xerostomia)?
> 3. How does calcium hydroxide produce dentin bridge?
> 4. Is there any way to replace lost tooth structure instead of using synthetic materials?

1.2.4.2 Applied Research

This problem-oriented research seeks to provide a practical solution to a specific problem, answer a direct question, and achieve the generally desired result quickly. The results can be used immediately after the project is completed. The purpose of this research is to put the practical application of knowledge into life. This research is more worthwhile, more objective, and more tangible. Therefore, in terms of efficiency, they have extraordinary value. In general, this research aims to improve the living conditions of human beings. It costs relatively less due to its short duration, and often a particular employer supports the costs incurred. However, the results may only be applicable at the time specified. These studies have the following characteristics:

1. In terms of time, they should be performed earlier than other types of research.
2. They are profitable.
3. Mostly public and private organizations and factories, and sometimes universities and research centers do this research.
4. The relationship between industry and academia follows this type of research.
5. Have a clear and generally professional employer.
6. Consider cross-sectional problems.
7. They are seriously looking for knowledge application in answering the problem, not the achievement of new knowledge.
8. Financial support is often guaranteed.
9. Most of their topics are about the validation or application of the results of basic research.

Examples of applied research in dentistry include the investigations to determine:

1. The immunomodulating properties of dental pulp stem cells.
2. The treatment options for aggressive periodontitis.
3. The ways to improve the antibacterial effects of root filling materials.

1.2.4.3 Developmental Research

Developmental research aims to improve and develop the previous devices, processes, systems, or existing situations. It is established upon applied research. For example, after identifying the cause of tooth caries in basic research and introducing possible solutions to treat caries in applied research, conducting a clinical trial to determine the best treatment or improve a current product is a form of developmental research. In many cases, applied and developmental research are considered the same.

Other examples of developmental research in dentistry are:

1. Improvement of an ultrasonic handpiece and its clinical evaluation.
2. Animal study for assessment of bone particles as implant material.

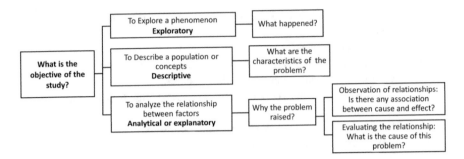

Fig. 1.5 Classification of research studies based on the main question of research (what the questions in each category are)

1.2.5   Based on the Main Question or Objective

Studies are based on different types of questions. Understanding this category of research is the foundation of any research program. The researchers must explain their aim clearly to be able to ask the correct question.

Research in this category can be divided into exploratory, descriptive, and analytical research (Fig. 1.5). Each of these types is structured in response to the main question of the study.

1.2.5.1   Exploratory Research

Exploratory research consists of describing or characterizing a phenomenon so that it appears familiar. It is, in fact, the starting point of research, with which the researcher intends to add to the existing literature in some way. From this perspective, it is perhaps the most explicit and most basic type of research. Being elementary does not necessarily mean that it is simple, but rather that the researcher takes the first step to build the foundation of a subject. Exploratory research seeks to answer the question of whether there is a problem in the research community. Examples:

- Are there any viable cells in the pulp of necrotic teeth?
- Are there any bacteria inside the disinfected root canal of a tooth?
- Does the amount of fluoride in drinking water vary in different parts of the city?

Obviously, in exploratory research, the researcher has no hypotheses, and his/her main goal is to gather information about the presence or absence of a phenomenon. Although the researcher has a presupposition in his/her mind, this assumption has no scientific basis. It is not permissible to interfere with one's presuppositions in the exploratory research process.

This research provides rich information about a disease. It makes the community aware of a problem that needs to be investigated in the next step on "how it arises" and "why" by conducting "descriptive" and "analytical" research. The researcher collects the data based on observations. Exploratory research can be classified into *case reports* and *case series*. A rare disease, a new symptom of a known disease, a new treatment, or the destructive effect of an intervention is reported in the case reports, and there is no control group. For example, you may aim to report on three cases of peri-implant bone loss and peri-implantitis [11]. *Case series* are based on reports of a series of cases of a given pathology or a series of treated patients without a specific control group assignment. In this type of study, we only consider a limited number of cases, so it should not be used for a risk assessment.

Here are the steps of exploratory research:

- Problem recognition: not clearly defined problem at a preliminary stage.
- Carrying out methods to explore the problem:
  - No hypothesis.
  - No control groups.
  - No intervention or manipulation of variables.
  - Produces qualitative data.
  - No conclusive result.
  - Leads to further research.

1.2.5.2 Descriptive Research

When the researcher accepts the exploratory research results and begins to describe the created phenomenon, he/she enters the descriptive research process. For example, after we find out that there are bacteria inside the disinfected root canal of a tooth, now we ask about their type, numbers, and properties. Therefore, if the research question is not about "what" a phenomenon is, but rather about its characteristics, that will be named descriptive research. In other words, we only describe the available information in descriptive research, which is part of the observational studies. The results of this research only show the features of a situation at the time of the study. For example, what is the prevalence or rate of caries in a population? What are the characteristics of this population (age, race, socioeconomic)? What are the radiographic or clinical symptoms?

Descriptive studies in epidemiology are often designed to express temporal, spatial, and disease distribution concerning individual risk factors and are examples of cross-sectional studies. The title of the descriptive research should be very detailed as follows:

- "Prevalence of oral diseases and oral-health-related quality of life in people with severe mental illness undertaking community-based psychiatric care" [12].

- "Prevalence of total and untreated dental caries among youth: the United States, 2015–2016" [13].

This type of research is a kind of *observational* study in which different characteristics of a sample of the population are observed. The questions such as "how many types," "what types," "how much," and "what ratio" are raised in descriptive research. These questions are usually answered via observation of the participants, interviews, surveys, obtaining measurements of physical features, or simply by extracting information from existing sources, such as disease reporting registers, hospital medical records, or employment service records.

Descriptive studies are cross-sectional surveys (opinion survey, knowledge, attitude, practice (KAP) survey), epidemiological descriptions of the occurrence of a disease, and studies of variations in health and disease trends as a function of time and territory. In *cross-sectional* surveys, cause-and-effect data are collected simultaneously, without determining the time sequence. They do not attempt to test a hypothesis about an association. They provide a prevalence rate at a point in time (point prevalence) or over a period (periodic prevalence). As the name suggests, a cross-sectional survey requires data collection on a cross-section of a population. This population could be the entire population of a city or a part (sample) of it. In this type of descriptive study, we use surveys to assess the distribution of a disease, pathological state, immunological condition, nutritional state, etc. The researcher always tries to report the facts without any interference. The goal in conducting such research is to provide a realistic description of the characteristics of a situation or subject. For example, you may interview a group of women to determine if they have children with cleft palate and whether they have taken certain medications during pregnancy. In such a study, you may see an association between medication use (exposure) and cleft palate (disease). However, the cause-and-effect relationship is unclear, as you did not select groups in this study.

The following example can better illustrate the steps of this study [14]:

- Study objectives: evaluate the occlusal, periodontal, and implant-prosthetic parameters and marginal bone loss (MBL) around implants after prosthetic loading.
- Identify all patients who received implants restored with single-tooth or up to three splinted crowns.
- Calculate MBL in all patients considering occlusal, periodontal, and implant-prosthetic parameters.

The critical point in this research is that, unlike exploratory research, the descriptive study is problem-oriented and hypothetical. The researcher must have a hypothesis, and his/her research will prove the main hypothesis, which is the answer to the main question. The analysis of descriptive data often leads to formulating a new hypothesis, which is the basis for analytical research.

The steps of descriptive research are:

- Recognizing the problem.
- Defining the hypothesis.
- Planning methods to answer the question.
  - No control groups.
  - No intervention.
  - Uncontrolled variables.
  - Produces quantitative data.
  - Conclusive result.

1.2.5.3 Analytical or Explanatory Research

While the exploratory and descriptive studies give an image of the current situation, the purpose of the *analytical research* is to go beyond and look for the links that events may have. In this type of study, the objective is to test hypotheses about differences between two or more groups or associations between various factors within the same group.

Analytical or explanatory research clarifies the relationships between phenomena and determines why such phenomena occur or under what conditions. In this type of research, the researcher seeks to answer the "WHY" question and discover the cause or determine the association between exposure to risk factors and disease. In other words, the objective of this type of research is to examine the causal relationship of variables. Exploratory research has allowed you to detect the presence of bacteria in the root canal of disinfected teeth; by descriptive research, you have known their characteristics; now you ask yourself why these bacteria can remain in the root canal even after disinfection. Before reaching this stage, the previous two steps must have been completed. Suppose the study's objective is to investigate the cause of the increase of dental caries among youth. In that case, the incidence of caries and their augmentation should be investigated earlier by exploratory and descriptive research. Accordingly, in this type of research, we need to know what others have done before. If you are sure about the increase of dental caries among youth and want to check the effect of immunoglobulin M (IgM) antibodies or the amount of fluoride in drinking water as causing factors, your study is analytical. In this study, the presence of a control group is mandatory, and the type of exposure to intervening variables is also essential (see the topic of variables).

The essential step in an analytical study is to develop a hypothesis and design a study plan that allows control of the variables that would interfere with the risk factor and disease. This approach varies depending on the analytical strategy adopted. Analytical research is divided into two types of observational and experimental (interventional) studies based on the assignment to exposure.

Observational analytical studies, which implicate only observing what is happening, do not involve any intervention. The researcher cannot control any of the third-party factors acting on any individuals under observation. In these studies, groups of individuals are compared to identify differences in exposure or outcome. Here, the observations of cause and effect are lagged in time and may give rise to an inference of associations. This association may even be statistically "significant." However, a high degree of relation between X and Y does not mean that X is the cause of Y or vice versa. You need an experimental study to verify causality. In observational analytical studies, samples are selected from a large group of people by random sampling; however, there is no randomization (random assignment of each subject to one of the groups). Therefore, groups have non-similar characteristics, and other factors besides the factor of interest may influence the results in these observational studies. *Observational analytical studies* could be prospective or retrospective, case-control, or cohort:

- In the *case-control* type, the case group has the disease of interest and the control group does not, and their history of exposure to the risk factor is determined. In other words, the groups are divided according to the dependent variable, and you will try to find out the possible cause of a disease. *Case-control* research is also called *retrospective* research, since cause and effect (independent and dependent variables) are examined after the occurrence. Here, you will determine the outcome before the determination of exposure to express "what happened" (Fig. 1.6). A higher risk factor frequency in disease cases than in controls indicates a relationship between it and the disease/pathology. The groups are selected based on the dependent variable. Matching the case and control groups is essential. Matching means choosing controls with similar characteristics (gender, race, age, etc.) to cases and is a method to reduce the effect of confounding factors. The matching ensures the features' comparability in the two groups; hence, a different distribution of these characteristics does not disturb the observed association between the risk factor and the disease. Selecting several control groups that represent pathological conditions other than the one studied would increase the power of the study. Exposure data is collected by researching the history and/

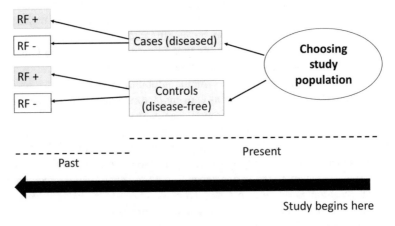

Fig. 1.6 Case-control study. RF+: exposed to the risk factor, RF-: not exposed to the risk factor

or consulting medical records through interviews, surveys, and/or file review. Sometimes the risk factor is a permanent condition, such as blood type, determined by clinical or laboratory examination. In this type of research, following a blind method is very important. The blind method means that the examiner or investigator should not know whether a subject belongs to the study group or the control group.

We can determine the associations between exposures and oral disease by using this study design. Suppose that you aim to "determine the relationship between amalgam restorations and periodontitis." Your analytical research method is case-control if you start with the selection of patients with periodontitis (case group) and healthy people (control group) [15]. Here, amalgam restoration is an independent variable (the exposures of interest), and periodontitis is a dependent variable. It is recommended that people in the case group be newly diagnosed with periodontitis and do not have any other oral disease. Recently diagnosed cases are preferred to eliminate the possibility of exposure to the risk factor studied after the onset of the disease in patients who have lived with the disease for a long time. Here, you determine the percentage of people in both groups who are exposed to an amalgam filling. Back to the example of children with cleft palate, if you choose two groups of mothers with healthy or cleft palate children and ask them about taking a particular medication during pregnancy, your research method is case-control. The following example can better illustrate the steps of this study [16]:

- Study objectives: investigate the role of alcohol consumption in implant failures.
- Matching criteria: age, gender, number of implants, the year implants were placed, bone augmentation.
- Identify all patients with implants with at least one failed implant removed or lost/exfoliated (case).
- Identify a control group of patients with the same characteristics and without any implant losses; select a random sample.
- Ask patients to report the amount of alcohol consumption per day.
- Calculate the frequency of alcohol consumption in each group and compare.

Therefore, the steps in the case-control research are as follows:

1. A clear statement of the inclusion, exclusion, and matching criteria to improve the validity of the results.
2. Selection of people with the disease of interest (case).
3. Selection of people without the disease of interest (control).
4. Measurement of the interested risk factor in the past of both groups (retrospective).
5. Data analysis to investigate the cause-and-effect relationship.

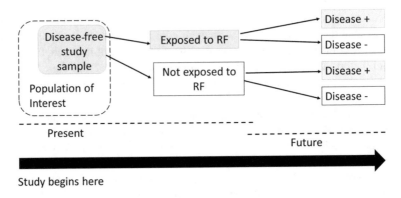

Study begins here

Fig. 1.7 Prospective cohort study. RF: the risk factor

- In *cohort* research, another type of observational analytical study that is more common than case-control in dental research, groups are divided according to the independent variable. The classic strategy of cohort studies is first to take a reference population without a given disease. Some individuals of this population present characteristics of interest for the study (exposed group), while others do not present them (unexposed group or control). Both groups should be free from the pathology considered at the start of the study. Observation of the two groups (with or without exposure) continues to identify the disease's incidence over a given period. Here, you will determine the outcome after the exposure or intervention to express "what will happen" (Fig. 1.7). In fact, in the cohort study, your goal is to show whether people exposed to a specific factor or exhibiting a particular characteristic have a greater risk of subsequently developing the disease than those not exposed or not exhibiting the particular feature. Different methods are used to collect the data, including investigations with questioning and follow-up procedures, medical records checked over time, medical examinations and laboratory tests, and the consolidation of archives with exposure datasets and outcome datasets. Let us consider a study to "determine the particle size ratio of composite fillers on the failure rate of restorations." The research method would be cohort if you select the teeth repaired with composites of different particle sizes and determine the condition of repairs (success or failure) after a while. In this case, the particle size is the independent variable (the factor of interest), and the failure rate (incidence) will be the dependent variable. In a cohort study for determining the relationship between amalgam restorations and periodontitis, you select a reference population without periodontitis. A group of this population has amalgam fillings, while the other one is without it. Then, you can assess the presence or absence of periodontitis disease in both groups by following them for a defined period.

    Regarding the relationship between cleft palate disease in children and the use of medication by the mother during pregnancy, if you select several children with or without cleft palate, and then ask their mothers about the use of the drug, the study is case-control type. In contrast, if you select a group of pregnant women, some of whom are on certain medications, and then examine whether their children will have cleft palate, your study will be a cohort study. When you plan a cohort study to evaluate the relationship between a population's oral health and mortality rate, oral health is an independent variable, and mortality is a dependent variable [17]. Another example of a cohort study could be the follow-up of healthy individuals with positive or negative gene polymorphism in specific interleukins and their examination at defined intervals for periodontal disease. In all these cases, subjects are disease-free at the start of the study, one of the groups is exposed to the factor of interest, and the number of new cases of disease (incidence) will be found out by follow-up time.

The following example can better illustrate the steps of this study [18]:

- Study objectives: investigate the effect of chronic periodontitis on the long-term implant prognosis.
- Matching criteria: type of implant, method of installation, suprastructure.
- Identify all patients with recent implant placement who lost their teeth due to chronic periodontitis.
- Identify patients with recent implant placement without a history of periodontitis.
- Compare the failure, success, and complication rates between two groups 1 and 10 years after implant placement.

The steps in the prospective cohort study are as follows:
1. The population of interest is determined.
2. Disease-free samples are selected in this population.
3. Some individuals in the study population were in contact with the risk factor (exposed), and some were not exposed to the risk factor.
4. At different time points, the occurrence or nonoccurrence of the disease is examined in each group.

Although in all the above examples for the cohort study, the research begins now, and the information is collected in the future (prospective), it is essential to note that the cohort study can also be retrospective. If the study design requires that you select individuals belonging to a specific population at a particular point in the past and then follow these individuals from that point in the past to "today," the study is called a retrospective cohort study. A retrospective/historical cohort study is only possible if the archives or data available make it possible to reconstruct a cohort exposed to a risk factor and monitor its effect over time. In other words, although the researcher did not attend the initial identification of the risk factor, he/she reconstructs the exposed and unexposed populations from the archives. Let us suppose you have access to the electronic records of adults at high caries risk, some of whom have received nonoperative anti-caries agents. You can perform a retrospective cohort study by determining whether they received anti-caries agents or not and calculating the caries outcomes in the data recorded from baseline to follow-up in both groups [19]. However, it may be challenging to prove that reducing tooth decay in all patients results from receiving anti-caries agents.

To better understand the difference between the retrospective and prospective cohort, take note of the following example:

- The goal of the study: Analyze the effect of implant location on the marginal bone loss around the dental implant.
- Prospective cohort study: Identify patients who were recently surgically treated with at least one implant; split them into two groups: implants located anteriorly and implants located posteriorly; monitor and examine them over a period; calculate the incidence rate for the development of bone defect at the site of implantation in each group and compare them (risk ratio).
- Retrospective cohort study: Using the records of a dental clinic, identify all the patients with implants placed 10 years ago; divide them into two groups: those with implants located anteriorly and those with implants located posteriorly; examine the data recorded during their follow-up regarding the crestal bone level to understand which patients have developed bone defect at the site of implantation and which have not; calculate the risk ratio in both groups.

The results of case-control studies and cohort studies may suggest a possible relationship between cause and effect. However, the best study design to prove a causal relationship is the *experimental (interventional)* study. The main characteristic of this type of research is manipulating at least one variable and monitoring and controlling other dependent variables to measure the effect of the manipulated variable (independent variable also called experimental variable or cause) on the monitored one (dependent variable). The general principles of experimental research are intervention, experimental group, control group, and randomization. This type of research can be viewed as the ultimate or decisive step in the research process, as a mechanism for confirming or rejecting the validity of ideas and assumptions about subjects' behavior or the effects on them. The researcher masters the subjects, the intervention, and the outcome measures, and he/she imposes the conditions under which the experiment takes place. More specifically, the researcher defines the subject who will be exposed and the one who will not be exposed to the intervention. This selection is made in such a way as to minimize any bias effects when comparing outcome measures between the exposed group and the unexposed group. In experimental (interventional) research, which is always prospective, the impact of an intervention on the living laboratory cells or bacteria, animals, or human beings is investigated before and after the intervention. If human intervention is performed, the study is called a *clinical trial*. Clinical trials are often used to test new dental materials, medication, treatment, or preventive regimens. However, any new materials or drugs should be evaluated via in vitro experiments and in vivo animal studies before entering the clinical phase. In the experimental group, intervention is the primary condition, and there is no intervention ("sham" intervention) in the control group. The control group can receive a placebo or a gold-standard intervention in clinical trials. Examples of experimental research include:

- Material: effect of thermocycling on dental composites.
- Cells: examining the impact of mechanical forces on osteogenic differentiation of dental pulp stem cells.
- Animal: studying the rate of gingivitis in amalgam-restored teeth in guinea pigs.
- Human: analyzing the effect of an herbal mouthwash on reducing dental caries in individuals with orthodontic appliances.

The following example can better illustrate the steps of this study [20]:

- Study objectives: correlate marginal bone loss around dental implants positioned subcrestally to short abutments.
- Enroll patients according to inclusion/exclusion criteria.
- Insert implants in the posterior mandible with the same protocol and deliver metal-ceramic restorations 3 months after implant insertion (adapt the length of the prosthetic abutments to the soft tissue vertical thickness).
- Evaluate bone loss after 6 and 12 months of functional loading.
- Calculate the association between prosthetic abutment height and bone loss.

A conventional clinical trial for a new therapy, device, or biomaterial goes through four phases:

1. Trials during phase I include studies involving limited healthy volunteers to evaluate the safety of a device. Giannakopoulos et al. conducted a phase 1 study for assessing "the cardiovascular effects and pharmacokinetics of intranasal tetracaine plus oxymetazoline" [21].
2. Phase II trials are designed to assess the clinical efficacy of the drug or device, determine the appropriate dosage, and study its safety. Additional pharmacological information will be collected, particularly on the dose-response relationship of the drug. If it is a device, its clinical efficacy should be evaluated, and its configuration and possibilities for improvement should be tested. An example is the study of Ciancio et al. for assessing the "safety and efficacy of a novel nasal spray for maxillary dental anesthesia" [22].
3. Phase III is the phase in which rigorous inclusion and exclusion criteria are followed for the choice of participants. The purpose of this phase is to assess the actual performance of a drug, device, or new treatment and to assess the safety of its prolonged use against the current standard treatment in a larger and more heterogeneous population than in phase II. This phase requires more clinical and epidemiological skills as well as high laboratory technology. Marketing authorizations for a new drug or medical device are based on evaluating the results of phase III trials. Therefore, phase III trials are subject to strict guidelines. Because phase III trials are time-limited, some side effects may not appear in the short term. Ciancio et al. conducted a phase III clinical trial "comparing three intranasal mists for anesthetizing maxillary teeth in adults" [23].
4. Although the market authorization of a drug or device is usually based on promising results from phase III trials, government agencies and the World Health Organization (WHO) prioritize using an additional test phase—testing in the field or under normal operating conditions. The purpose of the phase IV clinical

trial is to reassess the functional performance, safety, acceptability, and continuity of use of a drug or device under normal conditions of use. An example of the phase IV trial is the study of "oral dexmedetomidine premedication effect on preoperative cooperation and emergence delirium in children undergoing dental procedures" [24].

As in other research designs, it is rare for the researcher to study all the population in an experimental study. Therefore, you should select the samples representative of the populations from the target population to be analyzed for experimental purposes. A probability sampling generally does this. If the groups are very similar and the outcomes vary, the sample size should be more noticeable. It is also imperative to randomly assign subjects to the control and treatment groups. Randomizing the treatment allocation is very important in the randomized controlled trial (RCT) as the most powerful scientific research tool. There are many unknown factors, such as genetics or lifestyle, that can influence the results. Proper randomization can reduce the risk of severe imbalance of these factors in the various study groups. If the subjects in question are all available before the study begins, they can be randomly assigned into the control and experimental group after matching in terms of intervening variables (*randomized matched controlled trial (RMCT)*). Therefore, by random assignment in RCT, you can be sure about non-systematic differences between the treatment and control groups. Figure 1.8 shows the clinical trial design with random sampling and random assignment.

In most clinical trials, it is not possible to have the study group at the same time, and random assignment into the groups is not possible. In this *quasi-experimental* research, patients gradually come to the research site. It is necessary to gradually (non-randomly) assign individuals to intervention and control groups in such cases. Again, the groups must be matched in terms of intervention variables (group matched controlled trial). Suppose that you want to evaluate the efficacy of fluoride varnish on the primary dentition. If you randomly assign children to treatment and no treatment, your study is RCT [25]. In contrast, when children are part of an ongoing program, you cannot randomly assign them to treatment and control groups, so your study is quasi-experimental [26].

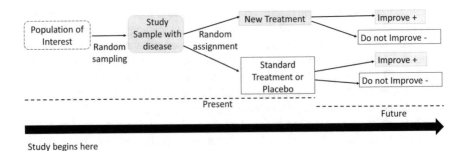

Fig. 1.8 Clinical trial design

Blinding is another crucial characteristic of clinical trials. A double-blind trial is a study where both the researcher team (including the statisticians responsible for analyzing the data and the investigators who write down the trial results) and the patient do not know whether the patient is in the experimental group or the control group. This method is advantageous when the control group receives a placebo drug or an identical sham procedure.

> The steps in the experimental study are as follows:
> 1. Identifying the relevant population of interest.
> 2. Determining the study sample by random sampling.
> 3. Obtaining "informed consent" from each of the participants before submitting them to the experiments.
> 4. Samples are then randomly assigned to the different intervention factors and the control group.
> 5. Subjects in the control and experimental groups are then followed up under rigorous conditions for a specified period.
> 6. Appropriate statistical methods are used to compare the outcome measures of groups.

The randomized split-mouth trial (a design of randomized controlled trial) is specific to research in dentistry. The split-mouth trial is defined as a study in which different sites in the oral cavity (tooth, quadrant, or jaw) of a patient's mouth will be randomly assigned to either the experimental treatment or control. The split-mouth experimental design has been used in many areas of:

- Dental surgery: "prophylaxis versus placebo treatment for infective and inflammatory complications of surgical third molar removal" [27].
- Periodontology: "antimicrobial photodynamic therapy as an adjunct to non-surgical treatment of aggressive periodontitis" [28].
- Orthodontics: "effect of micro-osteoperforation on the rate of canine retraction" [29].
- Cariology: "effect of sodium fluoride or titanium tetrafluoride varnish and solution on carious demineralization of enamel" [30].
- Pediatrics: "split-mouth design in pediatric dentistry clinical trials" [31].

In all analytical studies, the presence of a control group is mandatory, and the matching of the groups in terms of confounding factors, which distort the relationship between the exposure and the outcome, is essential, even though the process of randomization in clinical trials can remove these confounding factors. Before conducting the study, it is necessary to determine the inclusion and exclusion criteria, which always include the age range.

We have shown the classification of different studies based on the main question of research in Fig. 1.9. The process from simple observation to experimental research that follows a series of systematic steps is shown in Fig. 1.10. An observation (case report) or a series of observations (case series) launches a hypothesis. A

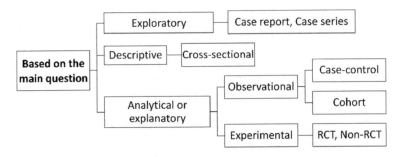

Fig. 1.9 Classification of research studies based on the main question of research (name of the studies in each category); RCT: randomized controlled trial

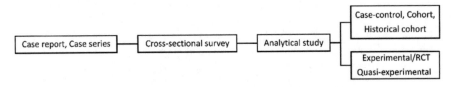

Fig. 1.10 The process from simple observation to carrying out experimental research

cross-sectional study (descriptive study) is undertaken to record the information on the characteristics of a disease or phenomenon, which provides a more in-depth view of the problem and generates a correct hypothesis. An observational analytical study including planned comparisons between variables gives more convincing evidence, establishes associations, and confirms (or rejects) the hypothesis. Finally, after this observation and these comparisons between groups, an experiment will be conducted to test the hypothesis more accurately and find the cause of the problem (experimental study). Let us consider that your colleagues in dental school report cases of implant failures. Cross-sectional research could address the prevalence of dental implant failures. By conducting a case-control or cohort study, you may find an association between bone resorption and the performance of dental implants. To confirm this association, you need to perform an experimental study. After realizing the cause of the dental implant failures, you may consider some solutions. Still, you need to do in vitro and in vivo studies to determine your intervention's efficacy.

1.3 Selecting the Research Topic

As mentioned earlier, research is a pathway to answer a question, and without a specific question, research will be aimless. Knowledge deficit within the research area leads to raising the research question. But what is the approach to get to the research question? In the first step, the researcher must determine the field of

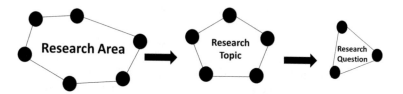

Fig. 1.11 The process of building a research problem that begins with moving from the research area to the research topic and then to a specific research question

interest or area of his/her research. Secondly, he/she must go through some steps to define the research topic, and finally, he/she can specify the research question that we will address in the next chapter. Your general "field of interest" will be the basis for choosing the "topic" of the research. Therefore, to start research, you must select the subject area of your research. Then, you need to define a research topic. For example, in the *field or research area* of "regenerative dentistry," your *research topic* could be about "the effect of laser on periodontal regeneration," "dental pulp stem cells for root canal therapy," "ceramic scaffolds for bone regeneration," or "biomimetic approach for dentin regeneration." After identification of the research area and the research topic, you should follow some steps to refine your "research topic" to a "precise research question" (Sect. 2.2). Therefore, building a research problem involves moving from an interest in a research area to a research topic and then to a specific research question (Fig. 1.11). The closer the researcher is to the research question, the smaller the dimensions of the problem.

1.3.1   How to Find the Field or Area of Interest?

The selection of the field or area of interest is the first step in conducting the research. Determining what you want to study (the subject area of research) and, subsequently, the topic for implementing any research project depends on many factors. The choice is influenced by considerations such as the possession of domain knowledge, your experiences, studies, needs, interests, and educational environment. Knowledge of current work lets you know if research is possible in the area and can suggest the type of question to ask and the precise subject to study. But how do researchers usually find the subject of their research?

The idea may come from personal experiences in everyday life, articles relating to the field of study, or research priorities of research centers.

1.3.1.1   By Accident or Personal Curiosity

Sometimes the choice of the research subject is entirely random. Luckily, the *subject* area of your project may be inspired by accident. For example, you notice that the number of referrals of children with molar tooth caries has increased. This increase in referrals will arouse your curiosity. Has there been a change in the fluoride of the drinking water in the area?

Sometimes paying attention to an ordinary situation can lead to a great discovery in science. For ages, people saw the apples fall from the tree. No one asked why objects always fall from the sky to the ground and not the other way around. But one person's attention to this seemingly commonplace issue led to discovering the law of gravity. Keep in mind that dentists have been using the "tissue engineering" approach a long time before its discovery by using calcium hydroxide as a pulp capping material. But no one paid much attention to the question, "how does calcium hydroxide induce the dentinal bridge formation?" Later, with the discovery of tissue engineering in medical science, the reality of this story became clear to dentists. The high alkaline pH of calcium hydroxide results in the release of growth factors from dentin, which subsequently lead to the recruitment and differentiation of dental pulp stem cells and regeneration of dentin (the formation of a dentinal bridge). For years scientists have known about the genetic programming of embryonic stem cells, which results in their multilineage differentiation into skin cells, neuron, muscle cells, etc. However, Yamanaka and Takahashi's curiosity to reprogram and reverse engineer adult cells into stem cells led to the Nobel Prize [32].

The critical point here is that the most beneficial research that will remain as the line of research and become the scientific heritage of a researcher is the one based on curiosity. Every dental researcher may deal with a variety of phenomena on a daily basis, which may stimulate his/her curiosity. The researcher should try to pursue a valuable issue that is vague to him/her for whatever reason, find something he/she can be passionate about, and be persistent enough to understand, clarify, resolve, or at least find a convincing answer for himself/herself.

The first sparks are ignited when you feel mentally challenged about a problem and want to find a solution. The researcher's personal interest is compulsory for choosing a subject area, and this interest is the engine of continuity of research. Identifying a knowledge gap or being fascinated by a problem can motivate the researcher. In this case, he/she will devote a lot of time and energy researching and overcoming limitations and challenges because his/her personal interest is at stake. Otherwise, the researcher will not be sufficiently motivated to continue the study in a desirable and valuable way and carry out the research to the end.

To find a field of research that attracts you, use the following questions:

- When dealing with dental patients or reading an article recently, did I encounter a topic that piqued my interest or made me think for hours?
- Is there an aspect of a subject in dentistry that I am interested in learning more about?
- Do I have a view on a current scientific controversy?

After selecting a research area, to understand if you are undoubtedly interested in that, ask the following questions:

- Why do I want to work on this subject?
- Will I still be interested in dealing with this subject in a few months, or even for years, as part of my career?

1.3.1.2 Previous Studies

One of the most important sources for finding a topic or subject is a previous study because it is always limited and specific. Each research project answers one or more questions but at the same time creates other questions or offers suggestions for further studies based on the limitations of the research. Therefore, previous studies can be used—as a source—to find new questions and choose a new topic. Keep in mind that in every research paper, some questions remain unanswered. Consequently, if you read the articles critically, you can find new subject areas or topics for your research. Usually, the last paragraphs of the "discussion section" of scientific papers are about the "study limits" and the "recommendations for future research." By reading these final paragraphs, you can determine the problems and questions that have not been addressed. Even if an article does not have a recommendation for future studies, by critically reviewing the existing literature and identifying the disagreements between researchers, you can think about what has not been studied yet or is missing in the literature. For example, in the studies on the effect of concentrated growth factor (CGF) on stem cell differentiation, if you notice that researchers have obtained different (controversial) results, you will be able to conduct a study that will close this debate by settling the question that remained unanswered. Systematic reviews that often focus on a particular topic are also excellent sources for finding new research subjects. They mention the remaining challenges after reviewing a wide range of research done on that topic. By reading articles, you may also decide to investigate the effect of other independent variables as causes of a dependent variable.

1.3.1.3 Research Priorities of Research Centers or Funding Centers

In many cases, organizations and educational or research centers submit a list of their research priorities annually, and budgets are allocated for this purpose. Depending on their field of study, the researchers can choose a subject area from the list, determine the topic, and receive a reliable funding grant for their research by successful approval of their proposal. For instance, the areas supported by the National Institute of Dental and Craniofacial Research (NIDCR) are "oral cancer, orofacial pain, tooth decay, periodontal disease, salivary gland dysfunction, craniofacial development and disorders, and the oral complications of systemic diseases" (https://www.nidcr.nih.gov/about-us/fast-facts).

1.3.2   How to Determine the Research Topic?

The research problem is an anomaly, an issue, a limitation, a difficulty, or clinical uncertainties that will be addressed through research. After choosing the area of interest or the field of research, the next step is to identify a problem that leads to the choice of the research topic. Let us suppose that you are interested in "dentin regeneration" as your field of research. You need to determine the subject of your research in this area based on a problem (dentin demineralization by caries) that must be resolved. If you understand the importance of a problem, you can properly articulate the problem and talk about the need to do a project for solving the problem. You can select the research topic, considering factors such as:

- Importance of the problem and its impact.
- Urgent need for a solution.
- Relevance to the objectives of the funding organization.
- Possibility of dealing with the problem through a study.
- Project feasibility.
- Chances of success.
- Foreseeable consequences in case of success.
- Benefits in terms of training and other elements strengthening research capacities.

To specify a research problem, we need to know "what is the process of selecting a research topic in dentistry?" Usually, the students are unaware of the importance of this process and haste to work on a research problem. Due to unfamiliarity with the process of selecting a research topic and lack of self-confidence, some students become afraid and anxious and decide to follow in the footsteps of other researchers and choose a topic that is not actually novel. The time you spend choosing a topic is critical.

Three essential steps to reach the research topic include observing, thinking, and reading about your area of interest. After explaining the first two steps in this section, the third step will be described in more detail in Sect. 1.4. These steps would help you to limit your "research area" gradually. Each step will pave the way for the next step (Fig. 1.12). The more accurate and systematic the previous step, the easier and quicker the next step will be. If you look closely at the problem, it will be easier to think about it. And if you have thought about the problem and considered all aspects of it, reading about it will not take much time. The more deeply the researcher observes the problem, the more systematically and creatively he/she will think

Fig. 1.12 The inverted pyramid of necessary steps of "observing," "thinking," and "reading" for determining the research topic

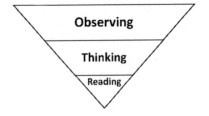

about it. The more time he/she spends thinking deeply, the more purposeful the reading will be and the sooner the goal will be achieved. Finally, allocating more time to *reading* will speed up formulating the research hypothesis and planning the research project. From this perspective, the three stages of *observing*, *thinking*, and *reading* are repeated regularly in the researcher's life, and each sequence in a new topic is based on the previous rounds.

1.3.2.1 Observing

Observation is an active process and the first step in gathering information about the research problem. When a patient comes to you and complains of a dental problem, you first receive information about the chief complaint and medical and dental history. Then by physical evaluation, clinical and radiographic examination, and the assessment of the performance of his/her existing restorations, you can have a complete view of the patient's condition. You try to answer some questions by detailed observation: Is there any vitamin deficiency in medical history? Has your patient had a new illness like diabetes? Are there any cervical abfractions, erosion on anterior teeth, or wear patterns on occlusal surfaces of teeth? Has there been a change in the patient's occlusion? Has he/she had bruxism recently due to stress? Has the patient developed new eating habits such as drinking acidic substances or smoking? How is the oral hygiene of your patient? Is there any infection?

In the same way, when you face a potential research problem, the first step is to gather information by detailed observation. You should actively observe as many details as possible about the subject to detect and assimilate any information. For example, one of your patients reports that he has stopped using dental floss because it tears when used for posterior teeth. If you are a curious person interested in the subject, the first step is to look closely at the situation. You must carefully examine the patient's teeth regarding too tight interproximal contacts, a marginal adaptation problem, or any overhang in proximal restorations. By thorough observation, you can collect some data about the situation.

Observation is not limited to the synthesis of new materials. Let us suppose that you are responsible for purchasing a composite for a dental school. The question is: Which brand of composite shows better performance in posterior teeth? This question can be the basis for starting a research project. The importance of this research is especially evident when thousands of composites are purchased annually for a dental school. Before deciding about the change of composite supplier, the first step is to gather the results of the colleagues' and students' observations on the composites' performance so far. How is the handling of existing composites? Are there any differences in terms of postoperative sensitivity, solubility, color stability, or wear rate? Based on the observations, you can move on to the next step.

1.3.2.2 Thinking

When we carefully observe a problem, we can then analyze it, understand its importance, and find solutions by creative thinking. Creative thinking is the ability to think in a unique way and beyond the established limits to arrive at authentic solutions. For innovation, we need to generate new ideas and new concepts through creative thinking on each question. As Albert Einstein said, "finding new questions or possibilities, and looking at old problems from a new angle requires a creative imagination." This facet allows us to innovate and face challenges in different ways and get out of the routine.

Generally, those who think more need to read and consult less, and those who think less seek to find the answer to their questions in the thoughts of others. Because creative thinking is not easy for many people, the shortcut is to get an answer from the product of others who have thought about it before. In this way, the researcher extracts his/her response from their works and puts it in front of his/her question. It seems that the precedence of reading over thinking saves time, but the disadvantage is that the imported answer is about another researcher's question.

Why don't we often spend a lot of time thinking about the research problem? Maybe we don't know what it means to *think*. It is widespread to confuse thinking with imagination. When you imagine that you have written your article and are now safely traveling and having fun or watching an exciting movie, you have not made an actual move; you are in the same position as before and only fly mentally in an unreal space. Thinking is a purposeful mental activity combined with developed imagination. The researcher should have a clear vision and manage it in the direction of the research topic. Because if you don't know where you are going, it will be impossible for you to reach your destination. For example, you are trying to figure out how or why dental floss tears in overhang areas. For this purpose, in addition to observation, you should think about the different aspects of it. What is the shape of the dental floss cross-section? Is it possible to change it? Is the dental floss single strand or multiple braided strands? Which one is better? What material is dental floss made of? Is it possible to use a different material? Is it possible to coat the current dental floss with a lubricant?

The more significant and more profound the initial thinking is on a topic, the other research steps, including reading and writing a proposal, will be easier. More practical thinking can be expected if you have a good background and experience in a research subject. On the other hand, the less knowledge you have about a topic, the less effective it will be to think about it, and the sooner you will have to move on to the next step (which is reading). In this case, as you did not spend a lot of time on the thinking step, reading will naturally take more time.

As mentioned before, it is not appropriate to prioritize reading before thinking; however, you can use brainstorming to get ideas from others to complete your vision. Teamwork is one of the main engines of creativity. By presenting the problem or the situation to colleagues, they could observe it, draw conclusions, share their ideas, and give you a new light on the case. Reframing the issue by others increases the number of angles from which you'll be able to consider the problem and discover new approaches. At this stage, accumulating knowledge is an opportunity to establish and to develop the research subject in mind and guess its various dimensions to explore the issue gradually.

Here are four critical tips in moving forward:

1. How to think is generally a matter of taste, and tastes are different, so the way of thinking comes back to us. Taking a bath, working in a noisy environment, exercising, and walking are examples of thinking methods that could be different in every person. However, imposing some limitations or deadlines could be beneficial for creative thinking.
2. Intellectual work must be accompanied by manual effort. Consequently, we must take notes while thinking. It's often helpful to use large notebooks with enough space to write a general description of various ideas. It is also beneficial to write each idea on a small card. Another way is to visually represent ideas in a diagram called the *mind map*. Usually, the main concept is in the center and sub-concepts radiate from the center. You can use keywords for preparing the mind map. That gives you an overview of a problem while helping to remember details. You can also link concepts to different branches. Using mind mapping software allows you to complete a mind map at any time. Take the time to develop several ideas through a mind map. You can highlight the potentials and weaknesses of each and create new ones if necessary. The words you have used during mind mapping can help you decide about your research methodology. If your keywords are more about the characteristics of a problem, you may choose *descriptive research*. In contrast, if you are more concerned about the relationship or association of several factors, your research methodology should be *analytical*.
3. The researcher must always be prepared for the emergence of new ideas. In other words, new ideas cannot be produced with prior intent. Therefore, it is recommended that the researcher always be prepared to deal with new ideas and thoughts and save them somewhere.
4. The researcher should try to generate new ideas without judging them. Avoiding being too rational can help you in this aspect.

The final question is how the researcher can discern the novelty and well-being of the thought that has arisen and formed in his/her mind. To answer this question, we can mention several well-known criteria:

1. Be objective: For this purpose, you should recall your goal of thinking and facts about the problem. Determine the problem (what is the problem?), its possible causes, and eventual consequences (why did it happen? what are its impacts?). Then, consider the compatibility of your ideas produced from your thinking with answers to these questions. You should not feel contradiction in your thoughts.
2. Ask questions, analyze the answers, challenge assumptions: The researcher must explain all the facts and dimensions of the subject in his/her mind and reflect well on their relationship with each other.

3. Be open-minded and communicate: Evaluate your thinking results by brain-storming to make its characteristics more transparent.
4. Think outside the box: To find an innovative topic, you need to consider its effect on others, not just how it affects you. Be concerned about at least one clinical relevance, research relevance, or educational relevance for your research topic. If there is no relevance (importance and usefulness), maybe you should reconsider your choice and find a new topic that is more engaging.
5. Identify missing information: At the end of this step, you should have prepared a list of questions for the next step, the literature review.

1.4   Handling Large Volume of Literatures as New Challenge in Performing Research

After observing a problem and understanding its importance, you start thinking about its various dimensions and then reading and collecting scientific evidence to increase your knowledge and ensure the choice of the topic (Fig. 1.13). Let's go

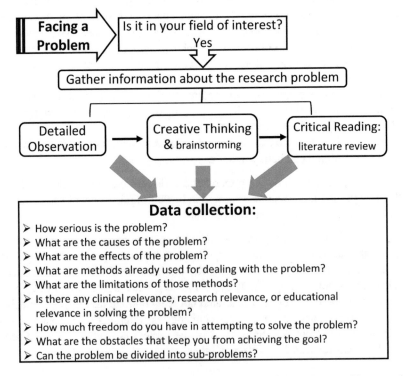

Fig. 1.13 The different steps of determining the research topic: from facing a problem to gathering the required information based on observation, thinking, and reading

back to the dental floss example. A researcher who has carefully observed and thought about the causes of floss rupture will now turn to read previously published articles to increase his/her knowledge. *Reading* at this stage means consulting the scientists who published on the related subject. The researcher should review the literature before starting the research design.

By *reading*, you can understand "what has been done on your research subject so far?", "how has it been done?", and "what are the results?". You will find out if there is enough information for your needs, set the perspective of your research, and have a more precise and more accurate topic. The more precise and purposeful the research topic is, the less time it will take to develop a coherent research design. Conversely, research design (Sect. 4.5) will be more time-consuming if the researcher spends less time reviewing the literature. We can briefly list the following advantages of literature review:

1. A better understanding of the research topic and increasing knowledge on the field of research: by examining the texts, the researcher expands his/her knowledge and becomes more informed about the subject. Consequently, the researcher can look at the problem with a broader perspective and begin to master the problem and refine its presentation.
2. Awareness of the research topic background: the researcher will be familiar with the scope and quality of the work that has been done so far on that topic.
3. Recognition of further research suggested by previous researchers or gaps in the literature might form the researcher's long-term goal and future research direction.
4. Learning how others have defined and measured key concepts allows the researcher to identify the desired variables in the study better, explain the cause-and-effect relationships between the variables in the form of theoretical models, and more easily formulate research hypotheses.
5. Assessing the appropriateness of the research topic and identifying the limitations and problems that other researchers have encountered while working on a similar issue will help the researcher complete his/her research with a clearer perspective, at a lower cost and greater efficiency.
6. Ensuring the innovation of the research question and its relevance to others: the researcher will not spend his/her time, energy, and budget on studying and examining the issues whose dimensions have been clarified already.
7. Finding people who are experts in this field and can communicate with them while working.
8. Saving time in the next steps of designing the research (preparing the proposal and writing the article): if the literature review is complete, the study's justification and rationale will be more straightforward in the next step (Sect. 4.5).

Innovation is essential in defining a research topic. During the literature review, if you are concluding that the problem has been dealt with before, you will no longer be able to work on it unless:

1. The type of research is different.
2. The main question or research hypothesis is different.

3. Your research involves new variables that were not previously evaluated.
4. A long time has passed since the previous research.
5. The previous study was done in a different place.
6. You aim to criticize the previous research.
7. You are anticipating comparing your idea with previous research.

Therefore, the literature review is an essential part of the research. At this step, reading articles should be quick so that the researcher becomes aware of the dimensions of his/her subject and can move on to another research subject if the issue is not novel. However, many books and articles exist on any research topic, especially if the problem is interdisciplinary and popular. Therefore, the researcher must know "what resources to choose for reading" and "how to read" them efficiently.

1.4.1 What Resources?

Thoughtful reading is a systematic and planned study. That's why when we have a subject in mind, we don't have to read all the books and articles related to our topic. By creative thinking about the research topic, we can use shortcuts in the reading phase. We have a list of questions from the previous step (creative thinking), and we only look for the answer to these questions from the resources.

When searching for information related to our topic, we must also consider the quality of the study. The literature includes academic articles and books that sometimes could be found on scientific websites like Scopus, PubMed, and ScienceDirect. Usually, books give you general information but may be out of date. "Peer reviewed" articles reviewed by two or three experts are good sources of detailed information. Also, multiple citations or continuous citations of an article or a book by known experts on a topic might be another criterion for using a source. High-impact journals or recognized scientists also can provide qualified articles for review.

We can access many articles in the first step—by entering a few keywords—even respecting the abovementioned criteria. Nevertheless, many of these articles have nothing to do with your core topic. The good news is that the structure of all articles is almost the same and includes the title, abstract, introduction, materials and methods, results, discussion, conclusion, and references. To find related articles, we must use the *elimination method*. First, we will skim the *title* of the articles. If the title is not related to our topic, we may remove it from our collection. The next step is to study the *abstract* of the remaining articles. If there is no clear objective, well-defined hypothesis, or precise conclusion based on the question in our mind, we can again skip the article. After skipping some papers in these two steps, we can search for the answer to our questions from the content of the remaining articles, which are the most relevant ones to our topic.

You want to know what these articles offer in relation to your research problem. Since you are at this stage in the preliminary step of research, there is no need to study the whole article. In fact, your topic should not be such that the entire paper needs to be read. If you feel that the entire article is related to the subject of your research, you must doubt the novelty of your topic because it is implausible that there exists such a detailed article on a new issue. Often, no article is entirely relevant to a new topic, but parts of the articles can be directly or indirectly related.

> We can summarize this section as follows:
>
> - Start by reading specialized academic books of prestigious scientists or articles of high-impact scientific journals.
> - If you have not reached a definite conclusion about your topic yet, start reading from more general books that briefly refer to your topic, and in the next step, after rethinking, go to the specialized books and articles.
> - If there are many new sources, do not read old-dated books unless the old date or old print is part of your research title.
> - For finding related articles, use the *elimination method*.
> - There is no need to study the whole article; just read parts that answer your questions listed before.

1.4.2 How to Read?

"How to read" does not mean having general literacy to read but having professional literacy and knowing the right way to review the literature. If we do not know how to read, we will be drowning in a sea of books and articles to the point of frustration, despair, boredom, and confusion, especially when we are on a tight timeline.

Many of us do not know how to read a collection of literature in line with our subject. On the other hand, 2.5 million new *scientific* articles are published each year. The researcher has a sea of recent articles in front of him/her and a collection of articles behind him/her. At the same time, he/she must also explore, innovate, and add to the existing knowledge. Most students ignore that having a lot of resources on a research topic does not necessarily mean that it is easy to do, but rather that it is challenging to study all these articles and extract new topics. So, what can we do? How can the researcher study all the related articles in a short period?

Just as there are particular methods and steps for preparing a cavity and filling it with restorative material, the literature review has a specific method that must be learned. We should have a plan for reading, break down our task into smaller components, recognize the necessary conditions for a good and valuable literature review, and believe that reading just any article is not necessarily beneficial. Let's consider the needs and stages of efficient and active reading:

1. One of the conditions for reading is choosing a proper environment and eliminating all distractions so that the researcher can appreciate "what he/she is doing." The researcher should study in a favorable environment to focus as much as possible on the study. If the external conditions of study are not available, intermittent, and/or cutoff issues catch the reader up, the time spent will not be adequate.

Everyone can concentrate in a different place, and "how we design the study space so that our focus is smoothed, and our study is not interrupted" depends on one's experience.

2. Unlike the previous condition, which was external, the second condition is more internal and is purposeful reading in line with the research plan and considering the question: "Why am I reviewing this article?" Accordingly, the intention (as an internal matter) must be set up in addition to the concentration (as an external matter). The researcher should prepare a list of questions related to the topic before overviewing the existing literature. This approach helps the researcher know precisely which parts of the article to go through, so he/she will not need to read the whole article. Preparing questions and asking them as you read will lead you to active reading and engage you in a conversation with the article. If we go back to the dental floss example, many questions arose from simple observation and thinking: What is the shape of the dental floss cross-section? Is it possible to change it? Is the dental floss single strand or a braid of multiple strands? Which one is better? What material is dental floss made of? Is it possible to use a different material? Is it possible to coat the current dental floss with a lubricant? How is the performance of dental floss measured? What type of control groups should be used? By asking these questions, you will have an active reading.

3. As mentioned earlier, the *title* and *abstract* are the first parts of an article that need to be read to determine whether you should read the entire article (Fig. 1.14). If the title is in the field of your research problem, you can read the abstract. By scanning the *abstract*, you will be able to answer questions about the article type and the author's goal (e.g., is the article a systematic review or original research? What is the objective of this research?).

   After reading the title and abstract, the first place to look at is the *results* section, especially the figures. The researcher should take a critical look at the "results" section of the article to determine whether the results are valid enough to use this article as a reference for starting new research. Here, you should pay attention to error bars on figures, confidence intervals, sample size, and control groups. By answering the question of "what the results show," you can draw your own conclusions from the article. Drawing different conclusions than those of the article's authors or not being able to conclude means that the paper is not a good reference for your research. If the results section seems interesting, you can consider reading the rest of the article.

   If the researcher has already prepared his/her questions, he/she knows where exactly he/she can find the answer to each question in the article. Suppose there

Fig. 1.14 Elimination method for reading the articles: scanning the title and abstract and then reading the content of the remaining articles

are questions like "what is the current knowledge about the research topic?, where are the existing knowledge gaps?, and how did the article fill the gaps?" in the list of questions you have already prepared. In that case, you must go through the "introduction." If the researcher does not have enough information about the background of the research topic, reading the "introduction" can provide information about the current knowledge of the research topic, the existing gaps in the literature, and the study's rationale. However, if you are familiar with the subject of your research, you can skip the "introduction."

In the "discussion" part, you can find more detail on the specific results and more articles related to your topic. At the same time, you should keep in mind that some sentences in the discussion are the author's own opinions and interpretations and may not be based on the experienced facts.

Finally, "materials and methods" gives you information about the procedures and equipment required to answer the research question. This section answers your questions about hypothesis testing techniques and methods, the study's population (humans, animals, materials, cells, or bacteria), variables, limitations, and the adequacy of the study's protocol in relation to the question asked. In the next step, by preparing a literature map, you can compare the methods used by different researchers who have worked on the same topic. It will help if you also consider whether the researcher has justified using the proposed method.

4. Having a critical mind is very important while reading the articles. Research is an activity of perpetual questioning which requires a skeptical mind, prepared for constant assessment and exploration. When instituted in a systematic framework, this questioning and this evaluation create the new knowledge necessary to carry out further actions for oral health.

5. Using software like *EndNote* or *Mendeley* can help you to collect and store the articles. As a critical reader, you can also use the software to comment on the references.

6. Taking notes during literature review is an essential condition of reading. The best way of taking notes is literature mapping, which can help you to see: "how does this article relate to other articles you are reviewing?" (Fig. 1.15). By doing that, you can find the resemblances, intersections, disagreements, and gaps in the literature, locate the subject of your study, and highlight important articles. Likewise, if you had some questions during the review of an article, the notes would help you to find the answer for them. Moreover, it can help you find where you need further study to collect more evidence and where there is no need to go into more detail. Various types of information can be distinguished in the literature map, and the relationship between ideas or outcomes can be articulated using circles, arrows, or different shapes. For structuring your map, you can consider different strands within your topic (patients, intervention, etc.), different backgrounds (methodological, geographical, etc.), development of ideas, or a combination of them. Assume that the general topic of your research is the effect of modified dentin on bone regeneration [33]. Figure 1.15 shows the literature map related to this topic.

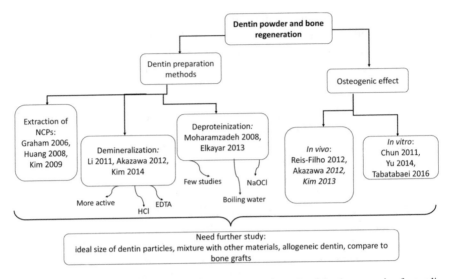

Fig. 1.15 literature mapping during a literature review on the topic of dentin processing for application in bone tissue engineering

References

1. F.J. Rodríguez-Lozano, S. López-García, D. García-Bernal, J.L. Sanz, A. Lozano, M.P. Pecci-Lloret, et al., Cytocompatibility and bioactive properties of the new dual-curing resin-modified calcium silicate-based material for vital pulp therapy. Clin. Oral Investig. **25**(8), 5009–5024 (2021)
2. E.B. Lubisich, T.J. Hilton, J.L. Ferracane, H.I. Pashova, B. Burton, Association between caries location and restorative material treatment provided. J. Dent. **39**(4), 302–308 (2011)
3. F. Alqudaihi, N. Cook, K. Diefenderfer, M. Bottino, J. Platt, Comparison of internal adaptation of bulk-fill and increment-fill resin composite materials. Oper. Dent. **44**(1), E32–E44 (2019)
4. R.P. Teles, V. Likhari, S.S. Socransky, A.D. Haffajee, Salivary cytokine levels in subjects with chronic periodontitis and in periodontally healthy individuals: A cross-sectional study. J. Periodontal Res. **44**(3), 411–417 (2009)
5. G. Laganà, N. Venza, A. Borzabadi-Farahani, F. Fabi, C. Danesi, P. Cozza, Dental anomalies: Prevalence and associations between them in a large sample of non-orthodontic subjects, a cross-sectional study. BMC Oral Health **17**(1), 62 (2017)
6. A. George, S. Ajwani, S. Bhole, H. Dahlen, J. Reath, A. Korda, et al., Knowledge, attitude and practises of dentists towards oral health care during pregnancy: A cross sectional survey in New South Wales, Australia. Aust. Dent. J. **62**(3), 301–310 (2017)
7. G. Fernandez de Grado, V. Ehlinger, E. Godeau, C. Arnaud, C. Nabet, N. Benkirane-Jessel, et al., Changes in tooth brushing frequency and its associated factors from 2006 to 2014 among French adolescents: Results from three repeated cross sectional HBSC studies. Denis F, editor. PLoS One **16**(3), e0249129 (2021)
8. E. Lempel, B.V. Lovász, E. Bihari, K. Krajczár, S. Jeges, Á. Tóth, et al., Long-term clinical evaluation of direct resin composite restorations in vital vs. endodontically treated posterior teeth — Retrospective study up to 13 years. Dent. Mater. **35**(9), 1308–1318 (2019)

9. D.H. Fine, K. Markowitz, D. Furgang, K. Fairlie, J. Ferrandiz, C. Nasri, et al., Macrophage inflammatory protein-1α: A salivary biomarker of bone loss in a longitudinal cohort study of children at risk for aggressive periodontal disease? J. Periodontol. **80**(1), 106–113 (2009)

10. M. Rajapurkar, A. Chi, B. Neville, B. Ogretmen, T. Day, PP081: Ceramide synthase isoforms in malignant transformation of oral mucosal dysplasia. Oral Oncol. **49**, S121–S122 (2013)

11. V. John, D. Shin, A. Marlow, Y. Hamada, Peri-implant bone loss and Peri-Implantitis: A report of three cases and review of the literature. Case Rep. Dent. **2016**, 1–8 (2016)

12. R. Patel, A. Gamboa, Prevalence of oral diseases and oral-health-related quality of life in people with severe mental illness undertaking community-based psychiatric care. Br. Dent. J. **213**(9), E16–E16 (2012)

13. E. Fleming, J. Afful, Prevalence of Total and untreated dental caries among youth: United States, 2015-2016. NCHS Data Brief [Internet]. **307**, 1–8 (2018)

14. C.D. Koller, T. Pereira-Cenci, N. Boscato, Parameters associated with marginal bone loss around implant after prosthetic loading. Braz. Dent. J. **27**(3), 292–297 (2016)

15. C.C. Ríos, J.I. Campiño, A. Posada-López, C. Rodríguez-Medina, J.E. Botero, Occlusal trauma is associated with periodontitis: A retrospective case-control study. J. Periodontol. **92**(12), 1788–1794 (2021)

16. R. Alissa, R.J. Oliver, Influence of prognostic risk indicators on osseointegrated dental implant failure: A matched case-control analysis. J. Oral Implantol. **38**(1), 51–61 (2012)

17. L. Jansson, H. Kalkali, N.F. Mulk, Mortality rate and oral health – A cohort study over 44 years in the county of Stockholm. Acta Odontol. Scand. **76**(4), 299–304 (2018)

18. I.K. Karoussis, G.E. Salvi, L.J.A. Heitz-Mayfield, U. Brägger, C.H.F. Hämmerle, N.P. Lang, Long-term implant prognosis in patients with and without a history of chronic periodontitis: A 10-year prospective cohort study of the ITI ® Dental Implant System. Clin. Oral Implants Res. **14**(3), 329–339 (2003)

19. B.W. Chaffee, J. Cheng, J.D.B. Featherstone, Non-operative anti-caries agents and dental caries increment among adults at high caries risk: A retrospective cohort study. BMC Oral Health **15**(1), 111 (2015)

20. T. Lombardi, F. Berton, S. Salgarello, E. Barbalonga, A. Rapani, F. Piovesana, et al., Factors influencing early marginal bone loss around dental implants positioned Subcrestally: A multicenter prospective clinical study. J. Clin. Med. **8**(8), 1168 (2019)

21. H. Giannakopoulos, L.M. Levin, J.C. Chou, A.T. Cacek, M. Hutcheson, S.A. Secreto, et al., The cardiovascular effects and pharmacokinetics of intranasal tetracaine plus oxymetazoline. J. Am. Dent. Assoc. **143**(8), 872–880 (2012)

22. S.G. Ciancio, M.C. Hutcheson, F. Ayoub, E.A. Pantera, C.T. Pantera, D.A. Garlapo, et al., Safety and efficacy of a novel nasal spray for maxillary dental anesthesia. J. Dent. Res. **92**(7_ suppl), S43–S48 (2013)

23. S.G. Ciancio, A.D. Marberger, F. Ayoub, D.A. Garlapo, E.A. Pantera, C.T. Pantera, et al., Comparison of 3 intranasal mists for anesthetizing maxillary teeth in adults. J. Am. Dent. Assoc. **147**(5), 339–347.e1 (2016)

24. S. Keles, O. Kocaturk, The effect of oral dexmedetomidine premedication on preoperative cooperation and emergence delirium in children undergoing dental procedures. Biomed. Res. Int. **2017**, 1–7 (2017)

25. H.P. Lawrence, D. Binguis, J. Douglas, L. McKeown, B. Switzer, R. Figueiredo, et al., A 2-year community-randomized controlled trial of fluoride varnish to prevent early childhood caries in Aboriginal children. Community Dent. Oral Epidemiol. **36**(6), 503–516 (2008)

26. P.M. Milgrom, O.K. Tut, L.A. Mancl, Topical iodine and fluoride varnish effectiveness in the primary dentition: A quasi-experimental study. J. Dent. Child. (Chic.) **78**(3), 143–147 (2011)

27. T.P. Bezerra, E.C. Studart-Soares, H.C. Scaparo, I.C. Pita-Neto, S.H.B. Batista, C.S.R. Fonteles, Prophylaxis versus placebo treatment for infective and inflammatory complications of surgical third molar removal: A split-mouth, double-blind, controlled, clinical trial with amoxicillin (500 mg). J. Oral Maxillofac. Surg. **69**(11), e333–e339 (2011)

28. A.L. Moreira, A.B. Novaes, M.F. Grisi, M. Taba, S.L. Souza, D.B. Palioto, et al., Antimicrobial photodynamic therapy as an adjunct to non-surgical treatment of aggressive periodontitis: A split-mouth randomized controlled trial. J. Periodontol. **86**(3), 376–386 (2015)
29. A.A. Aboalnaga, M.M. Salah Fayed, N.A. El-Ashmawi, S.A. Soliman, Effect of micro-osteoperforation on the rate of canine retraction: A split-mouth randomized controlled trial. Prog. Orthod. **20**(1), 21 (2019)
30. L.P. Comar, A. Wiegand, B.M. Moron, D. Rios, M.A.R. Buzalaf, W. Buchalla, et al., In situ effect of sodium fluoride or titanium tetrafluoride varnish and solution on carious demineralization of enamel. Eur. J. Oral Sci. **120**(4), 342–348 (2012)
31. A. Pozos-Guillén, D. Chavarría-Bolaños, A. Garrocho-Rangel, Split-mouth design in paediatric dentistry clinical trials. Eur. J. Paediatr. Dent. **18**(1), 61–65 (2017)
32. K. Takahashi, K. Tanabe, M. Ohnuki, M. Narita, T. Ichisaka, K. Tomoda, et al., Induction of pluripotent stem cells from adult human fibroblasts by defined factors. Cell **131**(5), 861–872 (2007)
33. F.S. Tabatabaei, S. Tatari, R. Samadi, K. Moharamzadeh, Different methods of dentin processing for application in bone tissue engineering: A systematic review. J Biomed Mat Res A. **104**, 2616–2627 (2016)

Chapter 2
Design Cycle of Research

2.1 What Is the Design Cycle?

Detailed observation of a problem, creative thinking, and critical reading about it generate a design cycle. When you come across a math problem, you *read* the *question carefully* several times (understanding the question is very important). In the next step, several solutions come to your mind to solve that problem. You usually try to choose the one that will get you to your destination faster. After that, you will solve the problem based on that solution and finally check it to ensure your answer. Consequently, "problem-solving" is a conscious, logical, thought-provoking, and purposeful process that helps you to find practical solutions to problems through it.

The selection of a research topic leads to the launch of a *design cycle* that is based on the problem-solving method. It is a series of steps that a researcher takes to start working on a project. The *design cycle* consists of four steps: (1) investigation (formulating the hypothesis), (2) designing a plan (writing the proposal), (3) performing the research (hypothesis testing), and (4) research evaluation (data analysis and interpretation) (Fig. 2.1). It is essential to stick to these steps because skipping one or doing them out of order will lead to improper results.

2.2 Investigation (Formulating the Hypothesis)

When we choose a topic for new research, creative thinking and critical reading of the existing literature and the current situation can give us insight into the prognosis of our study.

Now is the time to investigate and formulate the hypothesis, which includes four basic steps: (1) analysis of the collected information, (2) formulating the research question, (3) decision-making about the different ways to reach the answer, and (4) formulating the hypothesis. The research importance in following a design cycle

© Springer Nature Switzerland AG 2022
F. Tabatabaei, L. Tayebi, *Research Methods in Dentistry*,
https://doi.org/10.1007/978-3-030-98028-3_2

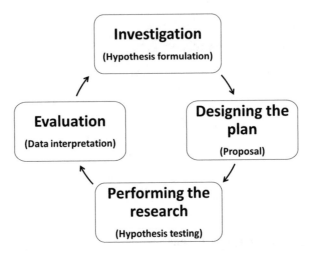

Fig. 2.1 The four steps of the design cycle of research

focuses on the "investigation phase," during which the researcher explores the problem and possible solutions. This investigation determines the overall direction of the research and should be very accurate; otherwise, the researcher will face many issues during the next steps of research.

2.2.1 Analysis of the Collected Information

After thorough observation, creative thinking, and critical reading about your research topic, you will have access to lots of information about the background of the problem and the methods already used for dealing with the problem. Now, it's time to examine the adequateness and completeness of the information collected from the previous steps. This step is crucial in the "investigation" phase. By analyzing the collected data, you can clarify the knowledge gaps, define the constraints that may affect your results while they cannot be changed, examine whether the problem can be broken down into smaller sub-problems, and consider potential solutions for each sub-problem. In addition to the scientific evidence gathered by literature review, other factors such as the researcher's experience and skills, as well as the research center's priorities, may influence the "data analysis" (Fig. 2.2). If the researcher doesn't have the required experience and skills to do the research, he/she should make sure that he/she can find an experienced member for their research team. He/she should also consider the required and available facilities (equipment and tools) for performing the research. In general, considering the researcher's skill and experience, research priorities of the university or research/funding center, and all scientific evidence in terms of novelty and solving a problem of society, the researcher should choose to work at the intersection of these three criteria.

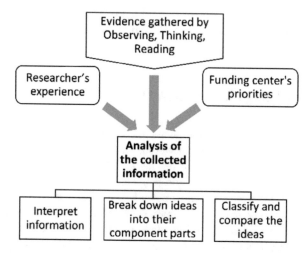

Fig. 2.2 The analysis of the collected information is not only based on the evidence gathered by observing, thinking, and reading but also the researcher's experience, as well as the research center's priorities

Analysis of collected information should enable you to move from the research topic to the conceptualization of the problem. To narrow down the subject of your research, you should:

- Define the problem (what is the problem?).
- Prioritize the problem (why is this a significant problem?).
- Find the rationale and relevance (can the problem be solved, and what are the benefits for society if the problem is solved?).

After a thorough analysis of the collected information, you may conclude that you need additional information, just like when, despite all the clinical and radiographic examinations of a patient, you decide that you need a blood test or consultation with other healthcare professionals. The final data analysis will give you a summary of relevant data and identify a specific problem with potential solutions.

Let us suppose that you have a patient with maxillofacial trauma. Careful observation shows that you must deal with multi-tissue loss, including bone and oral mucosa. You start thinking about the current treatments for these kinds of defects and their limitations. By reading articles, you realize that there are different options like autografts, allografts, xenografts, and custom-made 3D-printed plates for repairing these kinds of defects. However, all of them have some drawbacks, which led researchers to focus on bone and oral mucosa tissue engineering. You know that you need a tissue adhesive between soft and hard tissues for interfacial adhesion in this situation. Therefore, you will search about the required characteristics of an optimal tissue adhesive, the available tissue adhesives, and their limitations. Now that you collected a lot of information about the problem faced, it's time to analyze it. You can break down the problem into three sub-problems: (1) engineering oral mucosa, (2) engineering bone tissue, and (3) interfacial adhesion. For each component, you may have several ideas [1].

2.2.2 Formulating the Research Question

Formulation of a research question means interpreting and converting the research topic into a question that can be answered. Research questions are interrogative statements that develop and explain the identified problem. Just like a physician who examines all the physical and non-physical symptoms of a patient to make a correct diagnosis, the researcher must consider all aspects before formulating his/her question. By relying on the preliminary steps (observing, thinking, and reading), you will have a clear idea of the knowledge gap and the research problem. The results of the analysis step should now enable you to narrow down your research problem, formulate a single primary question explicitly, and define your objective.

The research topic will be formulated to a specific question that clearly defines the general objective. If you present the research topic without specifying the research question, it would not be apparent "how new knowledge will be generated." A well-defined research question is the essential ingredient of research. Being clear about the main question to be solved would increase the chances of success of the planned experiment, because an under- or over-specified problem can cause quite a bit of trouble later. The more precise is the question, the more likely the proposed research leads to new knowledge. It can also help the other researcher as this formulated question may answer their questions as well. Therefore, by asking answerable questions, you can provide new knowledge.

As Albert Einstein said, "the formulation of the problem is often more essential than its solution." Specifying the question makes it possible to give a framework to the project. The research question plays a guiding role because it will serve as a basis for formulating and identifying the hypotheses. Developing the research topic into a question makes it easier to outline the steps necessary to answer the question posed. It helps the researcher to understand more clearly what he/she is looking for. In other words, by framing the question, the problem is almost half solved. It is also helpful during proposal writing for answering the questions raised.

The type of question is wholly related to the type of research (Chap. 1; Sect. 1.2). If your study is exploratory research (*case report*), your question is about what happened (e.g., what are the adverse effects of tongue piercing? [2]). In the descriptive research (*cross-sectional* study), you will ask about the characteristics of the problem (What are the prevalence and risk indicators of peri-implantitis? [3]). In the observational analytical study (*case-control*, cohort *study*), you will ask about the relation of two or more variables together (Is there any association between orofacial pain and depression? [4]). Finally, the question in the interventional *(experimental)* research is about the effect of one variable on another one (Does preemptive analgesia with intravenous acetaminophen affect postoperative pain after third molar surgery? [5]).

A good research question has these features:

- Your question should be original (novel) and engaging. The originality of a research question means it has not yet been investigated from a certain angle you want to research. Therefore, a change of perspective can bring a specific form of originality to your research.
- The question should be pertinent to current scientific knowledge (educational relevance), to clinical and health policy (clinical relevance), or to future research (research relevance).
- The research question must be precise, coherent, and concise with a definite focus. Therefore, the research question should be neither long nor ambiguous so that the "who," "what," "where," and "how" of your idea can be recognized from your research question.
- A good research question is feasible. You may not be able to define a topic that has never been covered before. Feasibility also means that the number of subjects, the number of variables, and the technical skill for answering the question are adequate; it needs reasonable time and money, the concepts are clear, and the scope is manageable. You must ask yourself: "whether you have sufficient knowledge," "whether you have access to the study population," "whether you will have a large enough sample for your research," "whether you will be able to benefit from the help of other researchers," and "whether your research question is treatable in the allotted time and by available resources (financial means)."
- There should not be ethical issues in the research question or in answering it, resulting in the rejection by the institutional review board (IRB). It is therefore advisable to find out about the necessary authorizations before writing the proposal.

Therefore, you should formulate your research question while respecting novelty, clarity, feasibility, relevance, and ethical issues [6]. If you are studying the effects of photodynamic therapy on the microbial flora of the root canal, you might formulate the research question as follows:

- Does photodynamic therapy affect the microbial flora of the root canal?
- Or what are the effects of photodynamic therapy on the microbial flora of the root canal?
- Or could the maximum reduction of the microbial flora of the root canal be achieved after 40 seconds of photodynamic therapy?

However, the best way to formulate the research question is to consider four parts of the *PICO* (Patient/Population, Intervention, Comparison, Outcomes) framework [7] (Fig. 2.3):

1. P: the patients, populations, problem, subjects, or samples. For example, for evaluating the relationship between periodontitis and amalgam, you must define what group of society your question is about. People with a specific disease (patients with diabetes)? People in a particular service setting (dentists, dentist assistants)? Or a specific group of people (amalgam factory workers)?

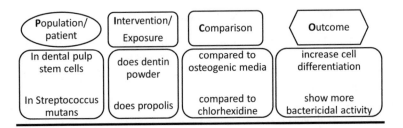

Fig. 2.3 The PICO framework for formulating the research question

2. I: the intervention (treatment, exposure, causative agent, diagnostic test). Here you should describe if you want to investigate exposure to a specific agent (capsules containing mercury and amalgamator) or a therapeutic or diagnostic intervention (amalgam filling/amalgam replacement).
3. C: the comparison intervention (no treatment, no exposure, alternative treatment) or the negative and positive control groups. You may also consider the gold standard as a control group for comparing groups. If the study group were patients with an amalgam filling, the control group could be persons without filling or having a composite filling.
4. O: the outcome or intended results. You should clarify what your intended outcome is. Is it curing, prevention, reducing death, reducing pain, recovery, the incidence of side effects, reducing the duration of hospitalization, change in the satisfaction rate, or improving the quality of life? Which outcome is to be affected? For example, the incidence or increase of periodontitis could be the outcome.
5. T: sometimes, you should also consider the factor of time like the age of patients, duration of intervention, or the time outcome will be reviewed.

Therefore, in our example about the effects of photodynamic therapy on the microbial flora of the root canal, the best question could be as follows [8]:

In endodontic retreatment of patients with peri-radicular lesions (P), does photodynamic therapy (I) have more effect than the 810 nm diode laser (C) in reducing the number of the microbial flora of the root canal (O)?

You can see two other examples in Fig. 2.3.

2.2.3   Decision-Making About the Different Solutions for the Research Question

Creative thinking and critical reading based on the research problem should lead you to several solutions that may resolve the issue, just like when several different solutions come to your mind to solve a math problem. Now, based on the existing conditions, it must be decided which solution is better. In the example of dental floss mentioned in Sect. 1.3.2, we may reach several solutions such as changing the shape of the dental floss cross-section, using multiple braided strands, changing the material that dental floss is made of, or coating the current dental floss with a lubricant. All the solutions for resolving the problem may be appropriate, but we need to choose the one that addresses the research problem most directly and holds the most promise of achieving the answer. The strategies for modeling the choices of solutions and the decision-making play an essential role in the effectiveness of the design cycle.

For making the best decision, the previous steps must be accomplished carefully and promptly. Based on the data of previous steps, different decision situations may be faced by the researcher:

- If there is sufficient information, the decision-making situation is *reliable*, and the result is predictable. When the problem is obvious, priorities are clear and consistent, and there is a well-defined research question, the researcher can make the right decision considering the time and cost-effectiveness.
- When there is not enough information about the research problem (lack of understanding or clarity about the critical parameters of a problem), the researcher is in an *ambiguous situation* and cannot make a good decision.
- Once there is a lack of information on some variables relevant to the problem, the researcher will be in an *uncertain situation*, and it isn't easy to predict the expected results.

It is imperative to take the time to assess in which direction you are going (case-control, cohort, clinical trial, animal study, in vitro study). You may further refine the details of required materials and methods, but you need to decide about your project's mandatory steps and general orientation. Any subsequent changes in the research type will cost you a lot of time and money. Based on your research question, you will have different options (Table 2.1). For example, if you want to check the prevalence of squamous cell carcinoma, cross-sectional studies are the best

Table 2.1 Decision about the type of research based on the research question

Question	Type of research
Epidemiology: Prevalence	Cross-sectional
Therapeutic:	
Effectiveness	RCT
Safety	RCT or cohort
Diagnostic:	
Validity	Cross-sectional
Effectiveness	RCT
Causality:	
Controllable phenomenon	RCT
Uncontrollable phenomenon	Cohort follow-up (exposed/unexposed)
Rare phenomenon	Case-control study
Prognosis:	
Frequent disease	Cohort (exposed/unexposed)
Rare disease	Case-control study

Fig. 2.4 The process of decision-making after formulating the research question

choices; if you're going to investigate the causes or the prognosis of this disease (uncontrollable phenomenon), cohort studies are appropriate; and in case your question is about the effectiveness of a new treatment for this disease, clinical trial studies are more applicable. Let us consider a research question on dental ceramics. If your research question is "which dental ceramics has the best fracture resistance among the available ones," you can measure bending (flexural) strength by in vitro assessment. However, if you want to evaluate the ultimate performance of a new dental ceramic, you need to conduct a clinical trial.

It is also essential to think about all constraints. You should carefully contemplate the solutions in terms of feasibility, safety, and ethical issues (Fig. 2.4). In this step, you should know everything about your study population, control group, and sampling. You need to consider if you want to evaluate the association between variables or the difference between groups. Another aspect is the number of observations collected per sample for each variable. Do you want to measure data in separate groups (unpaired) or more than once per same group (paired = different part of the same sample, split-mouth, before and after treatment)? What are your ethical issues based on the type of research?

Two fundamental factors in decision-making are the cost-effectiveness and time-effectiveness of solutions. You must plan your research based on the available budget, and you should finish your research at a specific time. In addition, in this competitive world where every day researchers present new study results as a published article, and companies offer a new product, it is essential to spend the least amount of time for solving a problem. Finally, if you want to translate your innovation to the clinic, you need to consider the patenting and Food and Drug Administration (FDA) approval/clearance procedure in this step [9]. You may need submission of a 510k notification for class II devices and dental implants or submission of a premarket approval (PMA) after conducting clinical trials for high-risk class III devices [10].

2.2.4   Formulating the Hypothesis

In the last step of the *investigation* phase, the research question must be translated into the hypothesis, and the researcher must imagine the anticipated answers to the question (Fig. 2.5). For example, your patients report increased halitosis and a decrease in toothbrushing frequency in 2020. You are looking to find the reasons for this, and you ask the question, "what is the reason for changing oral hygiene habits in 2020 compared to the previous years?" The first and foremost answer to this

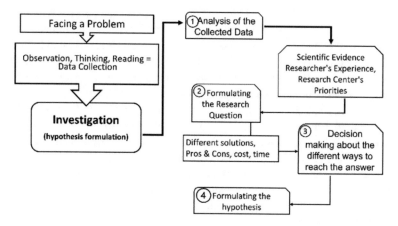

Fig. 2.5 Four basic steps of investigation phase of the design cycle to reach the hypothesis

question is the research hypothesis, for example, "the use of face masks during the COVID-19 pandemic impacts on oral hygiene habits and oral conditions" [11].

As we mentioned before, the *hypothesis* is a statement that predicts the response to the research question or a provisional declaration of the expected relationships between two or more variables. What will be the supposed behavior of a dependent variable after manipulation of the independent variable? It is the anticipated answer to the research question posed. Before performing the research, you will develop a prediction about the outcome of your study based on the logic of the problem. Then, by conducting experiments and collecting and analyzing data through research, you will test this hypothesis to see if your prediction is accurate or not. As Fred Kerlinger said, "hypothesis is perhaps the most powerful tool man has invented to achieve dependable knowledge."

The objective of a hypothesis is to provide a framework for answering the research question through statistical tests. The value of scientific work depends heavily on the formulated hypothesis. If a researcher has enough knowledge, he/she can make predictions by formulating hypotheses. It should be emphasized that hypotheses should not be lightly conjectured but must reflect the investigator's essential knowledge, imagination, and experience. In other words, the observation of "what already exists" and the literature analysis is the basis for developing the hypothesis. Its formulation must be precise so that the main objective of the project meets the identified problem. The hypothesis statement makes it possible to present the reasons or rationale for the experiment.

The factors to be considered for formulating the hypotheses are:

- The statement of relations: In most hypotheses, two main concepts are considered: causes or factors (independent variables) that affect dependent variables. The hypothesis expresses the expected relationship (association) between variables (independent and dependent variables) or the difference between groups (comparison). The relation between two variables can be causal (cause and effect, e.g., "this causes that," "this explains that," "this has an impact on that") or associative (e.g., "this has a connection with that," "this is related to that"). In a relationship between two variables, the variable to be explained is the dependent variable that you want to measure as your outcome. The explanatory factor is the independent variable that you want to change. The definition of the variable thus depends on the hypothesis specified in the study; a variable may be independent for one hypothesis, may be confusing for another, and may be dependent for a third hypothesis. For example, if your hypothesis is about "the relationship between smoking and periodontal disease" [12], smoking will be the independent variable. However, if the hypothesis is about "the effect of exercise on smoking," smoking is the dependent variable [13]. On the other hand, smoking could be a confounding variable in the "association between diabetes and periodontal disease" [14].

- If you specify the direction of the difference between groups in your expected outcome by terms such as "less than," "greater than," "positive," and "negative," you have a one-sided hypothesis. However, in a two-sided hypothesis, which is preferable, there is no specification about the expected direction of the difference. In this case, the meaning of the relation is indicated by terms such as "different from."
- Verifiability: The essence of a hypothesis is that it can be verified. It contains observable variables which can be measured and analyzed. Increasing the number of variables would increase the time required for data collection and complicate the study and the data analysis.
- Relevance: The hypothesis must be relevant to the phenomenon under study.

In scientific research, depending on the nature of the study and its specific aims, the number of hypotheses may be one or more. We can have a central hypothesis and secondary hypotheses. The number of hypotheses is not imperative, but their coherence and coordination are essential. Secondary hypotheses must be articulated around the central one and call each other in a logic imposed by the research problem.

When you formulate a research question based on the PICO elements, you can quickly acquire your hypothesis by changing it into a statement; by adding "not," you can then turn it into a null hypothesis (H_0). During the research process, the researcher's hypothetical answers to the research questions (H_0) are accepted or rejected (or the state of uncertainty remains unresolved, especially when the sample size is too small to ensure reliability). We can refuse the null hypothesis when the study results are statistically significant (a statistically significant difference between the variables). Answering a research question can lead to the emergence of a new question, constructing another null hypothesis, and continuity of the research. Consider the following example:

- Research question: In endodontic retreatment of patients with peri-radicular lesions, does photodynamic therapy in comparison to the 810 nm diode laser lead to a more or less reduction in the number of microbial flora of the root canal?

- Variables: Photodynamic therapy and diode laser are independent variables, and the number of the microbial flora of the root canal is the dependent variable.
- Hypothesis:

  - Directional or one-sided: In endodontic retreatment of patients with peri-radicular lesions, photodynamic therapy leads to *a more significant reduction* in the number of the microbial flora of the root canal than the 810 nm diode laser.
  - Nondirectional or two-sided: In endodontic retreatment of patients with peri-radicular lesions, *there is a difference* between photodynamic therapy and the 810 nm diode laser in reducing the number of the microbial flora of the root canal.

- Null hypothesis: In endodontic retreatment in patients with peri-radicular lesions, *no difference exists* between photodynamic therapy and the 810 nm diode laser in the number of the microbial flora of the root canal.
- The following null hypothesis after rejecting the first one: Different application times of photodynamic therapy have the same effect on reducing the number of the microbial flora of the root canal.
- Another null hypothesis: Maximum effect on reduction in the number of the microbial flora of the root canal cannot be achieved after 40 seconds of photodynamic therapy.

In the example of maxillofacial trauma mentioned in Sect. 2.2.1, you may consider the following research question and hypothesis:

- Research question: In multi-tissue loss, including bone and oral mucosa, can we adopt a new method of adhesion with suitable bonding strength that is biocompatible?
- Variables: Biocompatibility and bonding strength are dependent variables, and the adhesion method is the independent variable.
- Null hypothesis: In multi-tissue loss, including bone and oral mucosa, our new adhesion method is not different from the traditional method regarding bonding strength and biocompatibility.

Keep in mind that sometimes when you are researching and synthesizing an entirely new substance, you may not have a hypothesis about the properties of that substance.

The hypotheses are written in complete sentences and have a logical connection with the main research question and expression of the problem. The most important features of a reasonable hypothesis are:

- Clear, explicit, understandable, and feasible
- Specified before conducting the research
- A guide for the researcher that identifies the objectives and shows the general direction for conducting the research
- Consistent with existing knowledge: not against the facts and scientific principles of a field of knowledge
- Leads to the construction of a new hypothesis and further investigation

2.3   Designing a Plan (Writing the Proposal)

The Investigation provides evidence about the background of the research problem, the existing gaps, and the best way to bridge the gap. When you finish investigating the problem and the research hypothesis comes to mind, then you must write down how you intend to get the answer. Once the previous steps have been completed carefully, you know the significance and rationale of your research, what your aims are, how you want to test your hypothesis, your expected results, and your limitations and how they can be handled. Here you will define the problem, state the objective and specific aims, explain your study plan, and anticipate all the elements that will be useful and/or necessary during the hypothesis testing while considering the cost and time needed for each step (Sect. 2.4). In fact, by writing a proposal, you outline your research roadmap and show its logical steps from the start until the end. The proposal will serve as a guide in conducting, monitoring, and evaluating the research. If you want to conduct a survey, research questions, main objective and specific aims, relevance, methods of data collection and their justification, study population, sampling, list of resources, anticipated answers, and tools envisaged to evaluate data should be explained in the proposal. In short, everything that must be done for the project to succeed and achieve the goal must be presented and justified in the proposal.

Some components of research are beyond the scope of the *investigation* step. You should consider the success criteria for achieving your goal. You will also present the different constraints that you must consider before, during, or even after completing your project. Possible limitations during operation should also be considered. You should also explain who you are as a researcher and what your strengths are. Materials needed for the research and the cost of preparing or buying these materials should be identified at this stage, as well as the research equipment needed, the laboratories that have the equipment in question, and the cost of doing the work. Every study needs a budget, and you can only get the grant by writing a proposal. By preparing a proposal, you justify to the reviewers that you considered rationale, prognosis, and procedures, and you deserve to receive the grant (Fig. 2.6). In the fourth chapter of this book, we will explain how to write a proposal.

Not preparing a proposal and skipping this step can lead to many issues during research performing. The proposal helps you to focus on the requirements that have a significant impact on your results. It allows you to define your research team, materials and equipment needed, available facilities, and required budget. If you do not prepare the proposal, you may realize in the middle of work that you do not have access to the required material or that you cannot perform some of your tests at all. Therefore, proper research planning will help you start the research phase with peace of mind and without stress.

In a clinical situation, sometimes you repair the patient's teeth directly, whereas in other cases, you may use a provisional treatment that works as a pilot study for your treatment plan. Prototyping by provisional treatment can help you to discuss the required modifications with your patient and refine your plan. In a research project, the pilot study is a crucial part of a research design. It is an opportunity to

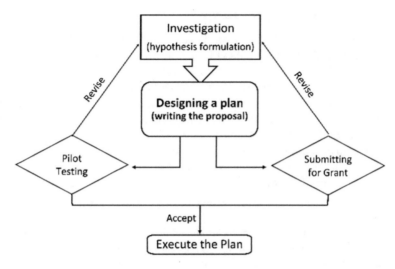

Fig. 2.6 Proposal writing: performing a pilot study and submitting the proposal would help to revise the plan and prepare it for execution

test and replicate design ideas and reveal the required modifications. The pilot study allows you to discover the weaknesses in design and instrumentation (external validity) and reconsider the sample size (internal validity), the consistency of the results, statistical analysis, research method, and budget. For example, to investigate "multi-level interventions at the practice and provider levels to increase dental utilization among 3- to 6-year-old Medicaid-enrolled children attending well-child visits," you need to present a formative and pilot work before the larger main trial [15].

2.4   Performing the Research (Hypothesis Testing)

Once the research problem has been sufficiently developed, the research plan has been written, and the necessary funding has been received, it is time to perform the research. The research team will test the hypothesis and collect data in a strategic network according to what has been planned in the previous step and the specific aims. The data collection process requires preliminary steps such as authorizations to conduct research, training of team members, managing the budget, strategies for data collection, management of potential problems, etc.

An experienced researcher, based on the literature review and the prepared proposal, knows well what techniques and tools he/she should use to measure the variables and collect the data. The types of tools can be:

- Interviews, questionnaires
- Measuring devices such as calibrated probes
- Films for X-ray images; radiographs
- Cellular/microbiological techniques

• Techniques for analyzing tissue samples

Organizations like the International Organization for Standardization (ISO) or American Dental Association (ADA) have defined several standardized methods and procedures for data collection, which routinely are used in dental research [16, 17]. Standardization is a process by which methodological differences between studies can be minimized. Therefore, whenever a standardized international recommendation for a procedure exists, it should be preferred and followed, and any deviation from it should be fully described and fully justified. You may have invented a specific way to conduct your research, in which case it must be reliable, valid, and practical, and not merely based on your own rules. These methods must be defined precisely to ensure that other researchers or critical reviewers can confirm your results or compare them with those obtained in the framework of other previously published studies.

Execution of the research design would help us to find out if our solution can resolve the problem. The more carefully and obsessively this step is done, the better the evaluation phase will be. Therefore, the researcher at this stage should apply the utmost care in performing the research method. The whole process of hypothesis testing must be written or traced by taking photos. That will help you in preparing the "Materials and Methods" section of your final report. During this step, if you made any changes in the proposal, you need to justify that.

2.5   Evaluation (Data Interpretation)

Once the research plan has been performed, you should evaluate your data based on the objective and success criteria that you have defined. By performing the research, you will have access to some raw data. To evaluate the collected data and find the answer, you must: (1) put these data in specific tables based on specific goals, (2) check whether the differences are significant or not by performing various statistical tests, and (3) finally interpret and discuss them. Therefore, the evaluation step consists of three phases: processing, analyzing, and interpreting data.

Processing means ordering, classifying, and grouping the data to be able to analyze them. Data should be isolated, grouped, and organized into categories, tables, charts, etc. By revealing the links that were not obvious, processing allows a large amount of information to make sense. It is, therefore, necessary to process data and transform them into analyzable data. Once you have your data files, do a final check to ensure they are correct before analyzing them. To verify the accuracy of data:

1. Watch for numbers that are not valid for a variable. For example, 140 GPa may not be a usual number for Young's modulus of cortical bone.
2. Make sure that the subtotals for the categories match the total number. For example, if you had 205 patients in your treatment group and the results show 85 of the patients were under the age of 21, and 130 were aged 21 or over, the total of these subgroups (215) does not equal to the total number of patients. This unmatching means that there are ten incorrect data you need to check.

For performing the second step, which is analyzing data, statistical software is employed purposefully to check the null hypothesis statistically. *Descriptive statistics* such as the mean, the median, or the standard deviation are applied to present data and should be summarized as mean ± standard deviation or mean ± standard error. Mean is the average of the values of a group, standard deviation (SD) is the variance or deviation of individual variables from the mean, and standard error (SE), which is preferable when you have a small sample size, is the variance of the mean. Calculated confidence intervals (CI), usually indicated as 95% CI and presented as a range, can be utilized to show the uncertainty of sample means. By employing suitable graphic representations, you can facilitate the visualization of data.

The descriptive statistic does not allow inference or prediction from the data collected. To draw conclusions, you need *inferential statistics* based on choosing an appropriate statistical test. To determine the proper type of statistical method (Fig. 2.7) that permits you to test hypotheses, you need to:

1. Specify if you want to compare means or variances (difference) or check the link between variables (correlation).
2. Review the distribution of each variable (you cannot use a parametric test for data without normal distribution).
3. Distinguish quantitative and qualitative data (Sect. 1.2.3). This distinction impacts how data is analyzed, as each must be evaluated and analyzed based on its own scales.
4. Identify the number of your groups (two groups or more than two).

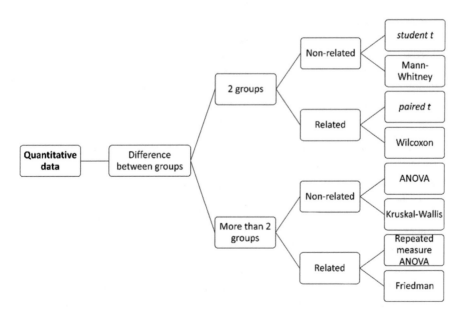

Fig. 2.7 Potential statistical analysis tests for quantitative data with normal or non-normal distribution

5. Specify if your data are un-paired/not related (different samples/subjects) or paired/related (taken from the same subject more than once, from different sides of the same tooth, various sites within the same oral cavity, or before and after treatment).

The *probability distribution* constitutes the crucial link between the population and its characteristics (variables) and lets us make inferences about the *population* by observing the *samples*. The probability distribution is a way of listing the range of values that a variable can take, showing the pattern of variation, and specifying the frequency of occurrence of each of these values in the population. The actual distribution of this frequency approximates a theoretical curve representing the probability distribution. Probability distributions are characterized by "parameters," quantities that permit us to determine the probability of a particular value of the variable. For proposing a statistical test, you should have good knowledge about *normal distribution*. By converting the dataset into a frequency (the number of specimens for different data ranges), you can realize if the data are normally distributed or not normal. *Normal distribution* means following a unimodal, symmetrical, or bell-shaped distribution on both sides around the mean. On the other hand, the distribution is not normal if you see a bi-modal or asymmetric distribution because of extreme values. You can check the normal distribution of data by statistical tests such as Kolmogorov-Smirnov and Shapiro-Wilk tests [18] or graphically using percentile-percentile (P-P) or quantile-quantile (Q-Q) plots [19].

Qualitative data, such as inflammation rate, pain level, microleakage degree, etc., which are assigned as "names/nouns" instead of values, can be nominal (like morphology of cells: columnar, cuboidal, squamous, spindle, stellate, or star-shaped) or ordinal with a natural ordering to the categories (like dental pulp irritation—slight, moderate, or severe inflammation). We can analyze the qualitative data by chi-square or Fisher's test. For example, in a study on clinical, radiographic, and histological outcomes after indirect pulp treatment of primary teeth, Fisher's exact test, Pearson's chi-square test, and McNamara's test were used for data analysis [20]. Qualitative data can also be converted to numerical values by scoring. For example, in examining the effect of different dental anesthetic techniques, the patient's pain can be numbered from 0 to 10, with 0 indicating no pain and 10 indicating intolerable pain [21]. However, converting qualitative data into numerical values is not possible in all studies. An example is when you study the effect of a substance on the morphology of cells, taking a microscopic picture of it and comparing the figures before and after treatment.

Mechanical strength, the solubility of materials, changes in salivary biochemical markers, etc., which are expressed by numerical values, are *quantitative data*. In the case of normal distribution, you can use *parametric tests*:

- A *Student's t*-test would be appropriate to compare quantitative data between two unpaired groups (not related), while for paired/related data, the *paired t*-test is more appropriate. For example, in a study on comparing two groups of nickel-titanium (NiTi) alloys (conventional and thermal-treated), the Student's t-test was used for analyzing data of torsional fatigue test (load applied until fracture), whereas paired t-test was applied for analyzing roughness (quantified in different parts of the instruments) [22].

- For comparing the means of quantitative data of three or more unrelated groups, the analysis of variance (ANOVA) would be more apt. When there is one dependent and one independent variable in three or more groups, the test would be one-way ANOVA (evaluation of cell proliferation in four different types of stem cells). Once you have one dependent variable and two independent variables, which should be compared in multiple groups, the test should be two-way ANOVA (interaction of age and gender factors on proliferation of different types of stem cells). If your data are related, *repeated-measures* ANOVA is more appropriate (the effect of time on cell proliferation of each group of stem cells by using a nondestructive method like PrestoBlue). After finding an overall significant difference (among all the means) via ANOVA, to find that this difference lies between which groups, post hoc tests such as Tukey (equal or more than five samples), Bonferroni (less than five samples), or Duncan (sample sizes differ between groups) could be practical. For example, in a study on the assessment of monomer elution from six bulk-fill and eight conventional resin composites ($n = 5$ in each group) in different storage media and different time points, comparisons between materials and monomer elution were evaluated by one-way ANOVA and Tukey post hoc test. Two-way ANOVA assessed the interaction between the type of composite and storage medium factors on monomer elution. The effect of time on the monomer elution was evaluated by repeated-measures ANOVA [23].

If the distribution is non-normal, the data should be analyzed by *nonparametric* or distribution-free tests such as *Mann-Whitney, Wilcoxon, Kruskal-Wallis,* or *Friedman*. We utilize *Mann-Whitney* when there are only two non-related groups, *Wilcoxon* for a paired two-group design, *Kruskal-Wallis* for more than two unpaired groups, and *Friedman* for paired multiple-group design. For example, in a study after covering dentin with three types of chelating pastes for three different time points, for evaluating dentin microhardness changes at four points of the dentin before and after treatment, the Kruskal-Wallis test was used for analyzing the differences between the different working times for each chelator (three sets of data). In cases of significance, the Mann-Whitney test was employed for pairwise comparisons. The Friedman test analyzed the differences between the microhardness of the different parts of the dentin (four points of dentin). The Wilcoxon test compared individual pairs of groups (before and after treatment) [24].

Example: When your research question is focused on comparing three groups (porous titanium implant and dense samples prepared by two different methods of 3D printing, and one commercial sample), your dependent variable could be the change in the mechanical strength (quantitative) due to the porosity. The independent variables will be the study group and method of 3D printing. You have one dependent variable and two independent variables, data are unpaired, and the distribution is normal. Consequently, you should use two-way ANOVA.

Table 2.2 Acceptance or rejection of null hypothesis based on type I and II errors

	The actual situation in the population	
Decision (based on the sample results)	H0 is true	H0 is false
H0 accepted	No error	Type II error (β)
H0 rejected	Type I error (α)	No error

If there is no statistically significant difference between the groups, you cannot reject the null hypothesis, while in the case of a significant difference between the groups, you can reject the null hypothesis. Two types of error are possible when you test the hypothesis (Table 2.2): In a type I error (α), you are wrongly rejecting the null hypothesis. For example, in comparing the effect of the photodynamic therapy and the 810 nm diode laser on the number of the microbial flora of the root canal, if there is no difference between them, but you conclude that they produced different effects, a type I error would occur. In a type II error (β), you are not rejecting the null hypothesis, while it is false. For example, in comparison of the effect of the photo-dynamic therapy and the 810 nm diode laser on the number of the microbial flora of the root canal, if there is a difference between them, but you conclude that they produced the same effects, a type II error would occur.

If the null hypothesis is not rejected, the test's statistical power ($1 - \beta$) must be calculated. When the statistical power is too low, it may be recommended to repeat the study with a larger sample size. If the power is acceptable, we can accept the null hypothesis. Sometimes, instead of deciding whether to accept or reject H0, we compare the statistical magnitude with the sampling distribution and calculate the value of α for which the test will reject the null hypothesis; it is called the p-value of the test. Remember that statistically significant results do not necessarily mean clinical significance. In the case of qualitative data, the total data must be analyzed and interpreted. In some cases, qualitative and quantitative data are complementary. It is sometimes necessary to collect additional data to refine the existing data.

The evaluation phase goes beyond data analysis, asks "why" and "how" questions, and offers interpretations. The researcher explains the data in the context of the study and the light of previous works after analyzing them. By verifying the validity of the results, going back on the hypotheses, and comparing the results with the data of the same type collected in the previous articles that approached the same question, the researcher will be able to make inferences, draw conclusions or develop a theory, and make recommendations.

The purpose of the evaluation phase is to see if the problem was resolved and if we have reached our specific aims. What makes us say that the goal has been reached? We will look at the achievement of the general objective concerning the expected results and other unexpected effects. Therefore, evaluation should be objective manner (Fig. 2.8).

- If you don't meet your general objective, you must go back and do a new design to understand why the aims were not achieved. In fact, by evaluating the research process, you look at the efforts made by your team members to perform the project and question the intervention strategy. You also ask how and why the methods

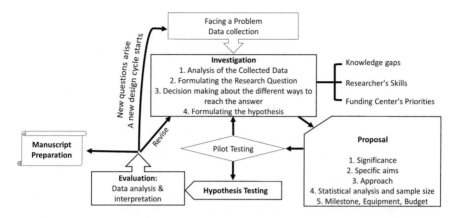

Fig. 2.8 A new presentation of the design cycle

do not work. This process would identify the potential constraints or the related elements that impact the methods. You can also determine the gaps between "what was planned" in the proposal and "what is implemented" in terms of resources, assays, and target groups. This step involves going back to the start of the design cycle and replaying each step to correct any errors that led to the failure to achieve the objective. Keep track of this work and write down your mistakes so that you don't repeat them next time.

- Once you have reached the objective, you can consider writing the final research report as an article or dissertation. You can introduce and present your research to your target audience and show "how you proceeded to obtain the results" and "how you could improve it" (Chap. 5).
- Finally, while interpreting the data and during this long process of your research, you must have reached a point where, as a researcher, you can make recommendations for improving the project, gaps to be filled, or new directions to be followed based on the limitations of your work. These recommendations should make it possible to improve the project and redirect it if necessary. From this perspective, many people consider research a helical model in which the end of each design cycle leads to new questions and suggestions for further studies and the start of a new design cycle. Keep in mind that performing valuable research that results in synthesizing a substance, submitting a patent, or even receiving a Nobel Prize is not impossible. Still, the key is to know that good research takes time, and you may have to repeat the design cycle frequently to get the desired answer finally.

The cycle of investigating, planning, executing, and evaluating the research is a process that helps to identify and analyze problems and limitations, formulate a hypothesis and test it, and finally evaluate the results to see if the problem was resolved. Following this orderly approach will increase your likelihood of success.

References

1. F. Tabatabaei, M. Rasoulianboroujeni, A. Yadegari, S. Tajik, K. Moharamzadeh, L. Tayebi, Osteo-mucosal engineered construct: In situ adhesion of hard-soft tissues. Mater. Sci. Eng. C **128**, 112255 (2021)
2. R. Shacham, A. Zaguri, H.Z. Librus, T. Bar, E. Eliav, O. Nahlieli, Tongue piercing and its adverse effects. Oral Surg. Oral Med. Oral Pathol. Oral Radiol. Endod. **95**(3), 274–276 (2003)
3. F. Vignoletti, G.L. Di Domenico, M. Di Martino, E. Montero, M. de Sanctis, Prevalence and risk indicators of peri-implantitis in a sample of university-based dental patients in Italy: A cross-sectional study. J. Clin. Periodontol. **46**(5), 597–605 (2019)
4. R.C. Castillo, S.T. Wegener, S.E. Heins, J.A. Haythornthwaite, E.J. MacKenzie, M.J. Bosse, Longitudinal relationships between anxiety, depression, and pain: Results from a two-year cohort study of lower extremity trauma patients. Pain **154**(12), 2860–2866 (2013)
5. K. Kano, K. Kawamura, T. Miyake, Effects of preemptive analgesia with intravenous acetaminophen on postoperative pain relief in patients undergoing third molar surgery: A prospective, single-blind, randomized controlled trial. Med. Oral Patol. Oral Cir. Bucal **26**, e64–e70 (2021)
6. S.B. Hulley, S.R. Cummings, W.S. Browner, D.G. Grady, T.B. Newman, *Designing Clinical Research*, 3rd edn. (Lippincott Williams & Wilkins, 2007)
7. W.S. Richardson, M.C. Wilson, J. Nishikawa, R.S.A. Hayward, The well-built clinical question: A key to evidence-based decisions. ACP J. Club **123**(3), A12–A13 (1995)
8. M. Asnaashari, M. Godiny, S. Azari-Marhabi, F.S. Tabatabaei, M. Barati, Comparison of the antibacterial effect of 810 nm diode laser and photodynamic therapy in reducing the microbial flora of root canal in endodontic retreatment in patients with periradicular lesions. J. Lasers Med. Sci. **7**(2), 99–104 (2016)
9. D.P. Taylor, M. Yoshida, K. Fuller, W.V. Giannobile, C.S. Sfeir, W.R. Wagner, et al., Translating dental, oral, and craniofacial regenerative medicine innovations to the clinic through interdisciplinary commercial translation architecture. J. Dent. Res. **100**(10), 1039–1046 (2021)
10. A. Purnama, D. Drago, *Medical Devices: FDA Regulatory Pathways for Medical Devices* (Regul Rapp, 2019)
11. C. Pinzan-Vercelino, K. Freitas, V. Girão, D. da Silva, R. Peloso, A. Pinzan, Does the use of face masks during the COVID-19 pandemic impact on oral hygiene habits, oral conditions, reasons to seek dental care and esthetic concerns? J. Clin. Exp. Dent. **13**, e369–e375 (2021)
12. J. Bergström, Tobacco smoking and risk for periodontal disease. J. Clin. Periodontol. **30**(2), 107–113 (2003)
13. B.H. Marcus, A.E. Albrecht, T.K. King, A.F. Parisi, B.M. Pinto, M. Roberts, et al., The efficacy of exercise as an aid for smoking cessation in women. Arch. Intern. Med. **159**(11), 1229 (1999)
14. A. Plessas, D.P. Robertson, P.J. Hodge, Radiographic bone loss in a Scottish non-smoking type 1 diabetes mellitus population: A bitewing radiographic study. J. Periodontol. **89**(9), 1043–1051 (2018)
15. S. Nelson, M.B. Slusar, S. Curtan, D. Selvaraj, A. Hertz, Formative and pilot study for an effectiveness-implementation hybrid cluster randomized trial to incorporate oral health activities into pediatric well-child visits. Dent. J. **8**(3), 101 (2020)
16. D.W. Jones, Dental standards: Fifty years of development. Br. Dent. J. **213**(6), 293–295 (2012)
17. B.W. Darvell, Misuse of ISO standards in dental materials research. Dent. Mater. **36**(12), 1493–1494 (2020)
18. M. Cierech, I. Osica, A. Kolenda, J. Wojnarowicz, D. Szmigiel, W. Łojkowski, et al., Mechanical and physicochemical properties of newly formed ZnO-PMMA nanocomposites for denture bases. Nanomaterials **8**(5), 305 (2018)
19. K.M. Hotchkiss, K.T. Sowers, R. Olivares-Navarrete, Novel in vitro comparative model of osteogenic and inflammatory cell response to dental implants. Dent. Mater. **35**(1), 176–184 (2019)

20. N. Sahin, S. Saygili, M. Akcay, Clinical, radiographic, and histological evaluation of three different pulp-capping materials in indirect pulp treatment of primary teeth: A randomized clinical trial. Clin. Oral Investig. **25**(6), 3945–3955 (2021)

21. P. de Camargo Smolarek, L.S. da Silva, P.R.D. Martins, K. da Cruz Hartman, M.C. Bortoluzzi, A.C.R. Chibinski, The influence of distinct techniques of local dental anesthesia in 9- to 12-year-old children: Randomized clinical trial on pain and anxiety. Clin. Oral Investig. **25**(6), 3831–3843 (2021)

22. E.J.N.L. Silva, J.F.N. Giraldes, C.O. de Lima, V.T.L. Vieira, C.N. Elias, H.S. Antunes, Influence of heat treatment on torsional resistance and surface roughness of nickel-titanium instruments. Int. Endod. J. **52**(11), 1645–1651 (2019)

23. R.Z. Alshali, N.A. Salim, R. Sung, J.D. Satterthwaite, N. Silikas, Analysis of long-term monomer elution from bulk-fill and conventional resin-composites using high performance liquid chromatography. Dent. Mater. **31**(12), 1587–1598 (2015)

24. M. Hulsmann, M. Heckendorff, F. Schafers, Comparative in-vitro evaluation of three chelator pastes. Int. Endod. J. **35**(8), 668–679 (2002)

Chapter 3
Systematic Review and Evidence-Based Research in Dentistry

3.1 What Is the Purpose and the Process?

The output of any research is new results or evidence that may be consistent or inconsistent with the previous ones. The assimilation of new evidence into dental clinics is essential; however, dentists usually have limited time to update their knowledge by reading all published studies. Most clinical dentists also lack sufficient professionalism to evaluate research methods, design, and the results of articles and cannot critically assess the validity and reliability of the published results. Subsequently, it is no surprise that a growing gap separates the community of dental researchers from those who can use the results of research in the clinic. The astonishing speed of the publication of dental research papers and the increasingly accelerated pace of therapeutic options and diagnostic tests have highlighted the need for evidence-based research in dentistry.

Evidence-based dentistry (EBD) based on the definition of ADA is "an approach to oral healthcare that requires the judicious integration of systematic assessments of clinically relevant scientific evidence, relating to the patient's oral and medical condition and history, with the dentist's clinical expertise and the patient's treatment needs and preferences" (http://ebd.ada.org). Evidence-based research in dentistry is a systematic process of researching and critically evaluating published evidence to determine the best treatment for each patient and to help advance basic clinical knowledge. Therefore, it aims to conduct high-quality research and appropriate validation that provides the best information for the dentist and proper treatment for the patient. It is a concept that has a profound implication for the practice and organization of clinical care. The objective is to implement treatments with a favorable benefit/risk ratio and collect information on their effectiveness and safety. It then integrates the most reliable results, clinical expertise, and patient preferences to guide decision-making about oral healthcare.

Designing a review article could also be the first part of a doctoral thesis which allows deeper expertise in a field. Research in dentistry is not like a stagnant

© Springer Nature Switzerland AG 2022 61
F. Tabatabaei, L. Tayebi, *Research Methods in Dentistry*,
https://doi.org/10.1007/978-3-030-98028-3_3

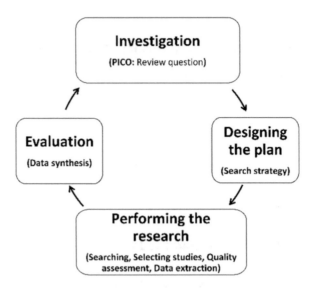

Fig. 3.1 Application of design cycle for systematic review

hermeneutic structure but like the form of a moving upward spiral, in which new answers lead to further questions and, consequently, the design of a new research project. As a result, systematic reviews can additionally illustrate the research's remaining questions and future direction in a particular field.

The design cycle described in Chap. 2 can help you perform a systematic review and find the evidence (Fig. 3.1). Let us suppose that your patient's tooth, after non-surgical root canal treatment, became reinfected. You have retreated the tooth non-surgically, but the disease persists. You want to know which option is best for your patient and has higher survival: tooth extraction and replacement using a single-tooth implant or intentional replantation [1]. In the *investigation stage*, using your knowledge, you can define a precise clinical question based upon which you can search for the answer. Given the existing clinical scenario for *designing the plan*, you should determine what type of study will better answer your question. Then specify your search keywords and inclusion and exclusion criteria. List all the kinds of information sources for yourself and write down the advantages and disadvantages of each. The statement of objectives and methodology should be explicit and reproducible. For *performing the research*, you should search in different databases (based on your search strategy), evaluate the articles considering the necessary criteria for the validity and reliability of the obtained studies, and extract data from selected papers. Finally, in the *evaluation* step, you can find the correct answer for your question by data synthesis.

Different types of clinical studies can be employed to assess the efficacy and safety of treatments. These studies provide results, some of which merit the name of *evidence*. A hierarchy of the types of studies has been proposed according to the level of evidence they provide. Among the different types of primary studies (clinical trials, cohorts, case-control, animal studies, and in vitro studies), the randomized controlled trial is the gold standard with the highest level of evidence [2]. Indeed, the randomized trial best minimizes significant biases, and it is the only one able to invoke the causal link between the experimental treatment and the observed results. However, we know that laboratory and animal studies are the foundation of clinical trials. On the other hand, the output of clinical trials leads to new questions that should still be answered through laboratory and animal studies. Another problem is the cost of clinical trials, and since dental treatments do not deal with the deaths of patients, the number of clinical trials in dentistry is limited. In other words, laboratory and animal studies could be accepted as a source of scientific evidence for clinical decision-making; however, a complete description of "materials and methods" is required in the published studies.

The primary purpose of evidence-based dentistry is to convert a complete set of research papers into one paper based on the best available evidence. It is recommended to develop a review protocol (like a research proposal) that includes the review's purpose, research question, search strategy, quality assessment process, data extraction strategies, and data synthesis in a defined timetable. Using a strict scientific design, you will have a *systematic review* by locating, evaluating, and synthesizing all the available literature related to a specified question. If you use quantitative methods for summarizing the results of articles, you will have a *meta-analysis*.

The method of finding the best evidence and doing a systematic review/meta-analysis consists of a rigorous methodological procedure as follows:

1. Defining a research question: The primary key to evidence-based research is to define a precise question. An accurate question will help you determine the right keywords for the search.
2. Specifying inclusion and exclusion criteria and determining explicit, transparent, and reproducible search strategy.
3. Systematic and exhaustive search, which attempts to identify all the studies meeting the selection criteria.
4. Evaluating the quality of articles systematically and critically: You should check the research's validity and relevance for critical evaluation. Keep in mind that any article published in a reputable journal will not necessarily be a good article.
5. Extracting data (details of any information) from selected papers and creating a comprehensive explanation of the best evidence available.
6. Data synthesis and showing the clinical relevance.

3.2 Formulating Review Questions

Today, evidence-based dentistry is accepted as a branch of dental research, and the need for an organized set of criteria for asking evidence-based dental research questions is recognized. The formulation of the review question is an essential step in the

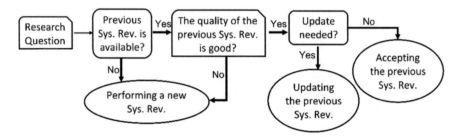

Fig. 3.2 The conditions showing whether a new systematic review (Sys. Rev.) is necessary or not

realization of systematic reviews. We will lose a lot of time and energy if we choose an incorrect research question. Therefore, to avoid this situation:

- The researcher should have enough knowledge about the fundamentals of research, expertise in the selected field, and awareness of any related reviews.
- The first step is to identify the need for a systematic review. For this purpose, you should begin your work by searching the scientific literature to verify whether a new systematic review is necessary or not (Fig. 3.2). Suppose no systematic review on the subject under study is available, or the identified systematic reviews do not cover all elements of a research question. In that case, a new systematic review should be performed. If a systematic review was produced a few years ago and recent relevant primary studies have been conducted since its publication, it must be updated.
- Sufficient primary studies ought to be done in your review question's field. The subject of the question to be researched should be imperative for the researchers and controversial between experts.
- The purpose of the review project must be defined: Is it a study intended for decision-makers to disseminate a medical technique? Is this a study designed for dentists or other researchers to help them in their daily practice?

Taking all these aspects into account will make it possible to formulate a precise research question.

The review question determines the theme of the systematic review and the main elements of the analytical framework. A well-formulated question will also guide you through the entire review's processes: search strategy, quality assessment, data extraction, and data synthesis. The question additionally specifies the dimensions to be addressed in the project, including the safety, effectiveness, and ethical, legal, or economic dimensions.

The research question of the systematic review should be formulated precisely and should be stated in observable or measurable terms. The question can be broad or narrow in scope. We should consider that if the research question is too broad, a narrative synthesis will be problematic due to the high number of studies that will need to be managed. On the other hand, a very narrow question could not provide a meaningful result as the number of studies will be too low. Its extent is a decision that depends on several factors, including the relevance and potential influence of the issue, the resources available, and the expected time frame. A well-formulated question considers all aspects on which the research needs to be focused. Suppose that your broad question is "Does regenerative dentistry work?". To achieve a formulated and focused question, you need to consider the following questions:

- What is "regenerative dentistry"?
- What kind of regeneration?
- In what contexts/settings?
- What does count as "regenerative dentistry"?
- What does "work" mean?
- What outcomes are relevant?
- Over what time periods?

The best way to formulate the research question in evidence-based research is to use the PICO framework explained in the previous chapter (Sect. 2.2.2). The PICO criteria clarify the question, specify the key concepts, and make an efficient and systematic bibliographic search in the databases. Thus, the PICO model facilitates the division of the question into different concepts, which will be used to construct the search strategy. Briefly, "P" stands for the patients or population (problem, subjects, or samples), "I" for interventions (the treatment, test, causative agent, or exposure), "C" for the comparison (control groups, no treatment, no exposure, or alternative treatment), and "O" is the output or outcome.

The *population* targeted by the main question should be representative of the population targeted by the outcomes of the data synthesis (last step of systematic review). The population can be defined in terms of disease or a problem of interest. It is also possible to describe or specify the population according to criteria such as age or sex if it is considered that these characteristics should be used to include or exclude certain groups from the systematic review. The restrictions must be justified and relevant. Depending on the project, it may be necessary not to exclude any group of subjects and instead perform subgroup analysis at the synthesis step. Factors to consider in defining participants are as follows:

- How is the disease or problem defined?
- What are the most important characteristics that describe the population?
- Should demographic characteristics be considered (e.g., age, gender, ethnicity)?
- Are there any groups of people/subjects that should be excluded?

Suppose you want to achieve evidence about the "diagnostic sensitivity and specificity of host-derived salivary biomarkers in periodontal disease amongst adults." In that case, you must define the periodontal disease and its essential characteristics based on the clinical examination. You should decide if you want to exclude participants taking medications or those with genetic or systemic conditions that affect periodontal disease [3].

The nature of the *intervention* of interest and the comparison, if any, should be clearly defined. The term "intervention" can refer to technology, a drug, or a mode of intervention. It is helpful to describe the intensity, frequency, and duration of the intervention as well as the context in which the intervention is carried out. The *comparison*, when applicable, must be well defined by indicating whether it is an inactive type (e.g., a placebo, no treatment, standard care) or an active type (e.g., a different intervention, gold-standard treatment). Factors to consider in defining interventions are as follows:

- What are the interventions of interest (exposure to a specific agent, a therapeutic, or diagnostic intervention) and comparison?
- What is the duration of the intervention and the duration of the follow-up?
- Are there variations in the interventions (e.g., duration, intensity, frequency, method of administration)?
- Should all variations of the intervention be included in the analysis?
- How will studies including only part of the intervention be considered?
- How will studies including the intervention of interest combined with other interventions (co-interventions) be considered?

For example, you want to evaluate the "efficacy of laser monotherapy or non-surgical mechanical instrumentation in the management of untreated periodontitis patients." Laser application is the *intervention*, and the *comparison* is non-surgical mechanical instrumentation. You may decide to accept all variations of the laser therapy (Er: YAG, diode, Nd: YAG with any duration, intensity, or frequency) while emphasizing the follow-up of at least 6 months. Finally, you may exclude studies including laser as adjunctive therapy or combined with supportive periodontal treatment [4].

The success or failure of an intervention is expressed in the form of *outcome*. The outcomes could be very diverse (mortality, health effects, quality of life, side effects, or economic results). In addition, authors of systematic reviews should consider the methods used to measure results by asking what type of scale or instruments were used and when the measurement was taken. If the list of desired outcomes is vast, they can be prioritized and categorized. Generally, you will consider the results that answer your foremost question. Factors to consider defining outcomes are as follows:

- List the expected results.
- Target the primary outcomes to be evaluated that would allow conclusions to be drawn about the main question.
- List the secondary outcomes.
- Specify the instrument for measuring the effects of the intervention and, if applicable, the time of measurement.
- Ensure that targeted outcomes examine all relevant results to answer the question.

For instance, for evaluating "the efficacy of subgingival instrumentation for the treatment of periodontitis," you may consider the reduction of probing pocket depth as the *primary outcome*, and number or proportion of pockets closed, changes in clinical attachment level, changes in bleeding on probing, or patient-reported outcome measures as the *secondary outcomes* [5].

Based on the PICO framework, the eligibility of articles (to be included or excluded from the review) will be determined. Therefore, if your topic is about the association of genetic polymorphism and external apical root resorption (EARR) [6], the PICO question will be: In patients undergoing orthodontic treatment, can genetic polymorphism alter a patient's susceptibility to EARR? In this example, *P* is "patients undergoing orthodontic treatment," *I* will be "genetic polymorphism," *C* could be excluded from the question, and *O* is "susceptibility to EARR." Here are some other examples based on the PICO:

"Are synthetic bone blocks effective for the treatment of bone defects with respect to other graft materials or ungrafted sites?" [7]

"Do bioactive dental implant surfaces have greater osseointegration capacity compared with conventional implant surfaces?" [8]

"Does applying mechanical forces change the in vitro behavior of stem cells in terms of proliferation/differentiation?" [9]

Evidence-based research questions can be in the field of diagnosis, treatment, prediction, or risk assessment. You can ask several different questions in each field, even if there is a previous review on that subject.

- Diagnosis: About the use of diagnostic tests. Example: How effective is DIAGNOdent in detecting proximal caries? Does it work better than bitewing radiography? Is DIAGNOdent effective in further patient cooperation? What are the success factors for better diagnosis of caries with DIAGNOdent? What is the ratio of risks to benefits of DIAGNOdent?
- Treatment: About providing the best treatment to the patient. Example: Do patients need antibiotic prophylaxis in dental implants and tooth extractions? What are the patients' needs for dental treatment after head and neck radiotherapy? What is the recall regimen and maintenance regimen of implant-supported restorations?
- Prediction: About estimating treatment or treatment problems. Example: What is the survival rate of an implant replacing a traumatized anterior tooth? Will this be different if the tooth loss is due to periodontal disease? What is the prevalence of caries in patients who drink fluoridated milk?
- Risk or danger: About the causes of diseases. Example: Does the posterior inlay increase the risk of postoperative sensitivity compared to other posterior restorations? What are the causes of periodontal (gum) disease in children?

Depending on the kind of question asked, the type of study selected would be different. If your question is in the field of treatment or intervention, it is better to find the answer by reviewing the randomized controlled trials (RCTs) and then cohort studies. But if the question is in the diagnostic field, cohort studies will give you a better answer. In the case of questions about prediction or risk, it will be more appropriate to examine the cohort studies and then the case-controls.

It is important to note that you can change your review question during the research, but you should know your chief motivation for changing it. Were you impressed by the results of some studies, or did you find that you were not initially familiar with all aspects and factors? By changing the question, your search strategy and the extraction method you defined in your protocol could no longer be acceptable.

3.3   Determining the Search Strategy (Documenting the Methodology)

The methodology of finding related articles for systematic review should be transparent and determined before starting the search. When your topic is complex or has more than one question, you can refine it by the search strategy. It also clarifies the assumptions of experts on the benefits and expected results of an intervention. Furthermore, it can show the context in which experts work or make decisions and highlight disagreements or controversies.

A good search strategy should be *sensitive* enough to avoid missing important information. It should also be so *specific* that there is no need to read thousands of articles to find those related to the criteria. Finally, it must be *systematic*, meaning so precise that others can repeat it if necessary. Applying more general keywords, the extended time period, more databases for finding articles, and using Boolean operators like OR and * in search strategy could increase its *sensitivity*. On the other hand, specific keywords, reducing the time period, particular databases, and using operators like AND and NOT would result in more *specificity* in the search strategy.

The *systematic* search process should be transparent, precise, and repeatable. Therefore, all the information related to search strategy, including keywords, name of the database, date run in the database, and limits or filters, must be recorded and reported in the manuscript. The steps for a *systematic* search strategy are as follows:

1. Based on the PICO format, select P, I, and O as your *keywords*.
2. Choose your database(s).
3. Decide to use different types of Booleans like AND, OR, and NOT, some limitations like type of study, language, the best period for including the published articles, or using the clinical queries in PubMed (Fig. 3.3).

   * Remember that the results of a good search should not exceed 300–400 articles, and if more, you should use filters that limit the results. By conducting a pilot test of your search strategy and testing your keywords and selection criteria, you may decide to adjust the process before executing the principal review.

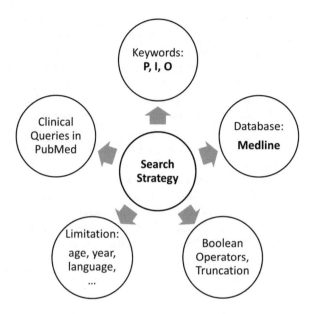

Fig. 3.3 Essential elements of the search strategy

3.3.1 *Choosing the Keywords*

The first step for determining the search strategy is analyzing the topic, selecting the keywords, and finding their synonyms, jargon, and Latin form. The selection of the keywords should be based on the question formulated in the PICO format. PICO elements can help you structure your search, but it is not necessary to use all of them. It may be preferable to avoid using the terms relating to outcomes if the database cannot identify relevant studies. Nevertheless, the type of study (ex.: RCTs) could be applied as one of your keywords. Therefore, you should choose essential keywords from your PICO format (P, I, O) and find out related or equivalent terms/ words by a generic search on the Internet. It may also be helpful to review the keywords used in relevant articles that have already been found and add them to the search strategy.

Here are the crucial tips for choosing keywords:

1. Consider at least three keywords for your search. Suppose that your PICO question is "Does the previous application of silver diamine fluoride influence the bond strength of glass ionomer cement and adhesive systems to dentin?" [10]. In that case, do not enter the exact phrase with prepositions and articles in the search bar, but only keywords such as fluoride (I), glass ionomer (P), and bond strength (O). It is essential to use the correct spelling of words. The search engine may suggest more convenient words.
2. Use noun, subject, or object as keywords. Keep in mind that entering verbs, adjectives, adverbs, and conjunctions will affect the search result.

3. Try to find synonyms, alternate spelling, related terms, truncations, and narrower or broader terms for your keywords. A drug may be mentioned by its generic name or its trade name. If your topic is "the effect of photobiomodulation on human dental pulp–derived stem cells" and your keywords are "I, photobiomodulation," "O, proliferation/differentiation," and "P, dental pulp stem cells," you can also consider "low-level laser therapy" for I, "dental pulp" for P, and "viability" for O [11]. There are also some sites for keyword suggestions like Google. com/trends, Ubersuggest.org, and Wordstream.com/keywords. However, it is better to use Medical Subject Headings (MeSH) for this purpose. If your keywords belong to MeSH, the exact English translation would be provided to broaden your research to international literature.
4. Use singular and plural versions of a word (example: enzyme, enzymes). This strategy will provide better coverage of articles related to your topic.
5. Enter the words that are most relevant in the search bar. For example, write "tensile strength" instead of "mechanical properties"; or instead of "impression," it is better to write "digital impression."
6. Databases are not sensitive to uppercase and lowercase letters. For example, the search results for "stem cells," "Stem Cells," and "STEM CELLS" will be the same.
7. Modify your keywords or their orders, if necessary, after starting the search.

3.3.2 Selecting Databases

Today, electronic databases such as MEDLINE, which are the predominant source of published articles, have made it easier to search for articles. One of the sayings of search engines is: "Stop searching; start Research," which means that you will save a lot of time to do your research by using the advanced feature of these engines.

The selection of databases to identify relevant studies for inclusion in the systematic review depends on the subject under investigation. The Medical Literature Analysis and Retrieval System Online (MEDLINE) database should be queried for health topics. MEDLINE is one of the best scientific databases covering various fields of medical sciences and is available to everyone online and free of charge. There are several ways to access MEDLINE, one of which is PubMed. PubMed (www.ncbi.nlm.nih.gov/PubMed) is designed by the National Center for Biotechnology Information (NCBI) at the National Library of America. Many full texts can be downloaded through this site or its link to the journal site.

Other databases that can be utilized are Google Scholar (https://scholar.google. com/), Scopus (https://www.scopus.com/), Web of Science (https://clarivate.com/ products/web-of-science/), Excerpta Medica dataBASE (EMBASE) (https://www. embase.com/), and Cochrane (http://www.cochrane.org/).

For some topics, searching the Internet will find more relevant data than searching databases. A search engine such as Google can be used to locate gray literature, conference reports, and abstracts. Keep in mind that public search engines, such as

Google, are not accurate in specialized topics and provide you with many articles and sites that may not be scientifically valid. The Google Scholar engine is more convenient for this purpose.

Information retrieval should combine several approaches to ensure that all studies relevant to the subject of the review are identified. To conduct a comprehensive search:

1. The search for articles should be done in at least two databases.
2. Seek studies and relevant information, which are not indexed in these databases, in other sources.
3. You can review the list of references of already identified articles to detect new studies that have not been found using the electronic tools.
4. Use the "Related citations" option in PubMed, the "Related articles" option in Google Scholar, or "cited by" in other databases.
5. Searching in Scopus can show you the ebb and flow of interest in a particular topic over the years and the experts who have worked mainly in this field.
6. Some search engines would list the articles that cited an already-found article. Consequently, you can also refer to those articles.
7. By clicking on an author's name, you may have access to other articles of this author that may be related to the one you have come across.
8. The data bank of clinical trials (https://clinicaltrials.gov/) and the International Clinical Trials Registry Platform (ICTRP) for health studies may help you to identify studies in progress and unpublished studies. Conferences proceedings, WHO reports, and letters to editors are other sources of unpublished data.
9. Moreover, you may consider checking special journals, thesis, and industry files or personally contacting scientists to find unpublished studies.

The following is an example of information retrieval from different databases in a study about the effectiveness of antibacterial monomers incorporated into dental adhesive systems [12]:

"This systematic review is described according to the Preferred Reporting Items for Systematic Reviews and Meta-Analyses (PRISMA) statement. Two independent reviewers (ARC and WLOR) carried out the literature search until 25th September of 2014. The following seven databases were screened: *MedLine (PubMed), Lilacs, Ibecs, Web of Science, Scopus, Scielo*, and *The Cochrane Library*. Moreover, the online system Questel Orbit (Paris, France) was accessed to recover patent documents related to antibacterial monomers in dental adhesive systems. Furthermore, a patent search was also made using International Patent Classification (IPC) with the following codes: A61K-6/00 (preparations to dentistry), A61C-13/23 (related to dental adhesive compositions), A61P-31/04 (related to antibacterial agents). The references cited in the articles included were also checked. After identifying articles in all databases, they were imported to the software Endnote X7 (Thompson Reuters, Philadelphia, PA, USA) to remove duplicates".

3.3.3 Boolean Searching and Selection Criteria

After defining the keywords (and their synonyms) and selecting the databases, it is time to determine the application of Boolean and the definition of inclusion and exclusion criteria.

It is better to start with a broad search with keywords, and then you can narrow down your research based on the results by applying advanced search, synonyms, Boolean searching (AND, OR, NOT), and some limitations. Your search strategy should be *sensitive* enough to find as many potentially relevant articles as possible. However, a large portion of the articles may not meet the inclusion criteria. You can increase the *specificity* of a search strategy and thus reduce the number of articles that are not relevant. You can retain certain types of studies (e.g., holding only randomized controlled trials) or limit the period you want to include articles in your research (from January 2015 to January 2022). However, greater specificity comes at the expense of sensitivity and may lead to the skipping of relevant studies. Therefore, the subject under investigation and the available resources must be in harmony with the articles identified by the search.

You can perform a more practical search via the advanced search feature of search engines and scientific databases. This way, you will find more sensitive, specialized, and relevant articles on the topic under discussion.

1. One of the ways to search for specific keywords and limit the results is phrase searching and putting them in quotations. This method would limit the results of your search. For example, compare this Google search: Dental materials = 253,000,000 results, "Dental materials" = 2,660,000 results.
2. Boolean operators allow you to combine several keywords/terms to refine or broaden your search.

   – Using the AND operator to search for two or more keywords (all of which would be displayed in the results) is a valuable way to perform specific searches and reduce the number of results. Almost all web search tools support this operator, although they may use other options such as "Must Include" or "all the words" to execute the AND operator. The AND evokes the constraint "must," limiting the search scope in the web environment. All keywords combined through this Boolean must be present in the search results (laser AND stem cells AND proliferation).
   – If you want to decrease the results further, using NOT or the minus sign (–: uppercase letter) is a good option. For example, suppose we wish to search for "implants." But we want to make sure that those referring to ceramic implants do not appear in the results list. Using this operator (implants NOT Zirconia) can limit the results.
   – By searching within the already retrieved result in some database (like SCOPUS), you can further narrow your search.

- Utilizing OR between synonyms is a suitable method to increase the number of search results and broaden the search (like "Dental plaque" OR "oral biofilm").
- Applying truncation or an asterisk after the first part of a word is also a convenient operation to find all terms that begin with that, including the singular and plural forms (gen*: includes "gene," "genes," "genetic"). However, it might also comprise terms like "generator," which are unrelated to your topic.
- Arrange the words in the order that is most important to you. To ensure that the search engine does the search in the order you want, place each word in parentheses (example: ("periodontitis") AND ("diabetes mellitus") AND (enzyme*) AND (treatment OR therapy)). By using the AND operator between the parentheses, you will narrow the search results.
- Usually, we use the quotation for two-word vocabulary, AND between main keywords, and OR between synonyms. Almost all search engines and scientific databases have an advanced search section that includes these features.
- You can record your search strategy by signing in to the search engines and turning on the "Search History" feature. After reaching the appropriate search strategy, activate the "search alert" option to automatically receive new articles in your email.

Figure 3.4 shows an example of using Boolean operators in a systematic review about the effect of concentrated growth factor on cells and tissues [13].

You should define your inclusion and exclusion criteria, including the type of study, gender, age, population, the similarity of exposure or treatment, the likeness of outcomes, minimum sample size or follow-up, etc. Study inclusion and exclusion criteria generally address different aspects of the research questions and state the type of studies targeted by the systematic review.

Inclusion criteria: The elements of the PICO will be employed to select the studies to be retained for the systematic review. You can also consider the research

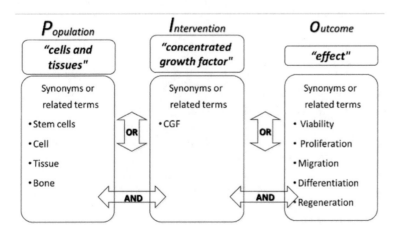

Fig. 3.4 An example of using Boolean operators to combine several keywords/terms

design and methodology. Similarly, you can adopt a specific research design (e.g., quantitative vs. qualitative; cross-sectional vs. cohort; observational studies vs. experimental studies), population (e.g., animal studies vs. human studies), sampling methodology (e.g., random clinical trial vs. non-random clinical trial), duration of treatment (e.g., long term vs. short term), or method of measurement in data collection. For a review question about "association of genetic polymorphism and external apical root resorption," you may consider some inclusion criteria like types of studies: which could be clinical trials, case-control studies, cross-sectional studies, or cohort studies. You can only select studies assessing EARR in orthodontically treated teeth as a primary object or studies in which genetic polymorphisms were quantified, and the existence of EARR was established [6].

For exclusion criteria, you can consider the type of study (cell culture, in vitro studies, or animal studies). You may choose geographic areas as an exclusion criterion for a review question on "caries development." You should also contemplate the age of patients evaluated in the articles or the minimum sample size or follow-up.

Employing methodological filters through Clinical Queries of PubMed is another strategy to select articles with specific research methods by emphasizing sensitivity (broad search) or specificity (narrow search).

You may limit your search by using the language (English) or the year of publication (from January 2011 to November 2021). Restriction of the time, language, or geographic considerations in some questions may result in a wrong answer. Therefore, narrowing the research question is better than using a lot of limitations to reduce the number of related articles.

The following is an example of inclusion and exclusion criteria of a study about the association of orthodontic tooth movement and external root resorption [14]:

"Inclusion criteria were: 1) population: human patients receiving orthodontic mechanotherapy; 2) intervention: orthodontic tooth movement with conventional fixed appliances or sequential thermoplastic aligners (STAs); 3) comparator: individuals or teeth within the same individual (including the split-mouth technique) not subjected to the same mechanical variable; 4) outcome: OIERR; and 5) study design: RCT, published or unpublished.

Exclusion criteria were: 1) studies of human patients with craniofacial abnormalities or nonhuman studies; 2) studies that include concomitant exogenous drug use, external stimulus, or adjunctive maxillofacial surgery; 3) studies including patients who previously received orthodontic treatment or combined surgical and orthodontic treatment; and 4) non-RCTs, cohort studies, case reports, case series, reviews, abstracts, systematic reviews, opinions, or studies where the diagnosis/measurement of OIERR was performed only on lateral cephalograms."

3.4 Performing the Review and Searching Studies by Using MeSH

Once your search strategy has been defined, you will be able to insert the keywords into the search fields of the various bibliographic search engines and start searching in selected databases. Always operate the "advanced search" feature of the scientific databases. During your search, you may decide to modify your keywords or search strategy based on the results (Fig. 3.5).

The process of searching for and retrieving articles should be repeated several times in the same database until you find no new articles. It is essential to update your search periodically, depending on the frequency of publication of relevant articles. You may also consider setting alerts in databases to receive new articles after completing your original search. This strategy ensures that the analysis will include the most recent publications on the subject.

As we said earlier, there are different databases for searching articles, but here we will focus on MeSH (Medical Subject Headings). MeSH stands for vocabulary used for indexing articles for PubMed. It was created by professional members of the National Library of Medicine to categorize similar titles in medical articles and libraries. MeSH vocabulary can help you use PubMed more accurately. All you must do is follow the step-by-step method below:

Step 1 Open the MeSH page (https://www.ncbi.nlm.nih.gov/mesh/).

Step 2 Assume that the subject of your research is: "Three-dimensional in vitro oral mucosa models of fungal and bacterial infections" [15]. You must first formulate the research question using the PICO format. In the example above, "P" is oral mucosa models; "I" is infections; and "O" is the survival. Usually, if you enter a word in the search box on the MeSH page and click on the search field, a list of all the words that the MeSH has matched for your target word will be displayed. The MeSH also provides a scientific definition for each word that defines the meaning of that word. Start with the word "oral mucosa." The results show that the MeSH term for "oral mucosa" is "mouth mucosa" with entry terms: Mucosa, Mouth; Oral Mucosa; Mucosa, Oral; Buccal Mucosa. If this word does not seem appropriate to you, you can choose another term from the list of linked terms at the bottom of the page (example: Mucous Membrane). You can try one of the subheadings related to your search (example: microbiology). Click on the "add to search builder," the term "Mouth Mucosa/microbiology" [Mesh] will appear in the "PubMed Search Builder"

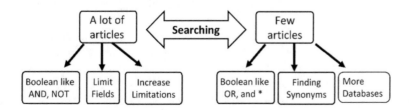

Fig. 3.5 Modification of keywords or search strategy based on the results

box. Decide about your Boolean (AND, OR, NOT) before going to the next step. By doing this, you specify for PubMed that your keywords will be more than one item. Now you can use another word from the "PICO" category, find the related words from the MeSH database, and add the appropriate words to the previous ones. The more keywords you send to the search box, the more specific your search will be, and the more irrelevant articles will be eliminated.

Step 3 Once you have added all your keywords, you can ask PubMed to find the articles you are looking for based on what you have sent to the search box. To do this, press the "search PubMed" button. The list of articles that appears can be further narrowed by using MY NCBI FILTERS (publication date, article type, language, sex, age, subjects, etc.). You can save, email, or send the result of your search to *My Bibliography* or *citation manager* and create an alert for future published articles in this area of research. You will have access to *My Bibliography* if you sign in to NCBI or register for an NCBI account.

Finding required articles is a crucial step in developing a systematic review. It aims to identify all scientific publications relevant to the research question, necessary for understanding and analyzing the subject under study. Web research should be conducted transparently in a structured manner. The information-gathering process should be described explicitly in the final report of the systematic review. Detailed presentation of the information retrieval process allows other researchers to replicate it or the project team to redo it for updating the systematic review. The comprehensive demonstration of the process also allows the reader to assess the quality and robustness of the information retrieval performed in the review context. A summary should be inserted in the "Methodology" section of the final report.

The following is an example of the search terms used in a study about the relation between periodontitis and peri-implant diseases [16]:
"The present study searched for articles published until March 2018 without date restrictions. The databases Medline via PubMed, and Cochrane Library were searched as follows: ((peri-implant disease OR peri-implant disease OR peri-implantitis [Mesh] OR periimplantitis [Mesh] OR peri-implant mucositis OR peri-implant mucositis) AND (periodontal disease [Mesh] OR periodontitis [Mesh] OR chronic periodontitis [Mesh] OR aggressive periodontitis [Mesh] OR risk factors [Mesh])). The Web of Science was searched as follows: ((peri-implant disease OR peri-implant disease OR peri-implantitis OR periimplantitis OR peri-implant mucositis OR peri-implant mucositis) AND (periodontal disease OR periodontitis OR chronic periodontitis OR aggressive periodontitis OR risk factors)). A manual search was also performed in the list of references of the included studies in attempt to find items not found in the electronic search (Fig. 1). Grey literature was searched on the U.S National Institute for Health (U.S. Clinical Trials) using combined uni-terms of "peri-implantitis" and "periodontitis" or "periodontal disease". The references were organized using the Reference Manager® software version 12.0.3 (Reference Manager, version 12.0.3, Thomson Reuters, Philadelphia, USA)."

3.5   Quality Assessment

Assessing the quality of articles refers to the critical assessment of the validity of studies. You can draw valid conclusions about the effect of an intervention as long as the research methods and the results of the included studies are accurate. Quality assessment in the systematic review is based on pre-established criteria. Since the quality criteria for studies may vary depending on the discipline or the field of application, it is essential to specify in advance what these criteria will be.

As mentioned in Chap. 1, you will access many articles by entering keywords in the search engine bar; however, many of these articles have nothing to do with your core topic. First, you need to combine all the search-strategy results applied to different databases and remove duplicate records. Then, you should screen them by scanning the articles' "title" and "abstract" to find related articles based on the selection criteria. At this stage, if the title/abstract is not related to your topic, you can remove the paper from your collection. Suppose that you have found 1000 articles from three different databases by applying your search strategy; after reviewing the title and abstract, you may exclude 800 articles for 1 or 2 reasons. Finally, you can check the full text of the remaining articles, which are the most relevant ones to your topic, and critically evaluate their content. This step may result in the elimination of more articles based on your pre-established criteria for quality assessment.

Contact the authors, if necessary, before deciding on whether to include an article in your systematic review or not. The information available in the article may not be sufficient to determine a decision. By contacting the authors, you may obtain more information. Then, you can decide on the admissibility of the article and record the responses received. If the authors do not respond within the time limit set, you can exclude the study in question, indicating for exclusion: "potentially relevant studies, but it is impossible to rule on the criteria for selection due to incomplete information."

Decisions on whether to include or exclude studies should be documented and detailed in a flow diagram that shows (Fig. 3.6):

- The number of studies identified in each database according to the established search strategy
- The number of publications identified from other sources (list of references, citations, etc.) (including the number of duplicates)
- The number of publications retained or excluded after the first scan of the "titles" and "abstracts" and the reasons for exclusion
- The number of publications retained or excluded after reading the full texts and the reasons for their exclusion

To evaluate the quality of the included studies, you should have more than one assessor (at least two) who works individually for appraising the compliance of articles with your criteria. This approach reduces the risk of excluding relevant studies, minimizes the risk of error in judgment and subjectivity, and ensures the reproducibility of results. The decisions made at each step by each reviewer should be

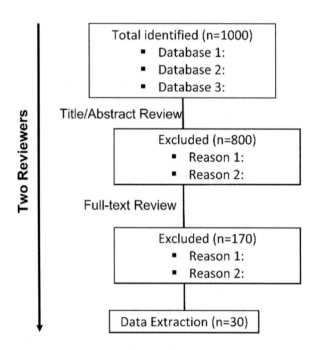

Fig. 3.6 The process of selecting studies for including them in the systematic review

documented to verify the degree of agreement between them subsequently. Disagreements between two reviewers should be resolved by discussion or by the presence of a third party. You can also use Cohen's kappa coefficient (k) to evaluate the two reviewers' search concordance. The process of resolving a disagreement between reviewers in selecting studies should be specified in the protocol before-hand and included in the final report.

Usually, different aspects of the research questions would be considered for quality assessment. Again, one of the best ways to evaluate articles is to use the PICO format for each article:

– P: Who are the source population or samples, the eligible population, and the participants investigated in the article?
– I: What is the intervention used in the article?
– C: What groups were compared?
– O: What outcomes were assessed? Is there any problem with data reporting?

By critically assessing articles and going through them sentence by sentence, you may find a lot of problems in them, causing their exclusion from your list. You may use a personal checklist for evaluating articles:

– Who are the authors of the article?
– What is the credibility of the authors or the reputation of the journal? To which university or research center are the authors affiliated?
– When was the article published? How many citations until now?

- What is the purpose of this article? Is the purpose clearly stated?
- Do the goals of the study and its design contribute to understanding a situation or clinical decision? What is the relation of the work to other pieces of literature in the field (diversity in the type of study, methodology of research, or statistical analysis could inhibit the achievement of a good answer for the question of systematic review)?
- How is the study done? Is the method used for conducting the research appropriate? Is there any alternative for the method applied?
- Has a proper comparison been made with the control group?
- What are the sampling methods, data collection methods, ethical issues, and data analysis methods?
- Is the method of performing the assays described in full detail so that it is possible to repeat it?
- What are the results of the study? Are the results presented clearly? Do these results help to provide better treatment for patients? Do the results of the study seem valid?
- How are the studied population compared to actual patients?
- In addition to the statistical significance of the results, is the clinical significance also considered? What is the scientific value of research?

For increasing standardization in examining the methodological quality of the studies, several evaluation tools have been developed. These tools come in the form of a *checklist* or a rating *scale*. The *checklist* asks a series of questions without providing an individual or summative value, while a rating *scale* assigns individual values, or scores, to a series of questions and groups them into a summative value. The gathered information through the assessment tools can qualitatively assess the articles by categorizing them as high, intermediate, or low quality. As a result of this evaluation, it will be possible to exclude studies of lower quality or consider the quality level of the studies in the following steps of data synthesis and conclusions drawing. It may be appropriate to exclude studies of lower quality when many studies of good quality are available. On the other hand, when few articles are available or studies are methodologically very heterogeneous, it is recommended not to exclude based on quality criteria.

The choice of a tool depends on the type of research of the articles, its relevance to the review's objective, and its own validity. To ensure proper methodology, you can apply the following tools:

- The PRISMA (preferred reported items for systematic reviews and meta-analysis) statement (http://www.prisma-statement.org/), consisting of 27-item checklist, has been prepared to improve the quality of review articles. This program is a list of criteria that should be considered in a systematic review.
- Clinical trial evaluation standards were also established and adopted by CONSORT (consolidated standards of reporting trials).
- The Jadad scale was developed to provide quantitative scales of sample randomization, blind, and subject dropout among the significant issues in clinical trials. The sum of scores for these three issues ranges from 0 to 5. A score of 2–0 shows the poor quality of the article, while a score of 3–5 indicates its high quality.

- The Newcastle-Ottawa Scale (NOS) can be applied for evaluating the quality of nonrandomized clinical trials.
- Cochrane collaboration tool is more beneficial to examine the potential of risk bias in randomized controlled trials.
- Animal Research: Reporting of In Vivo Experiments (ARRIVE) with Systematic Review Centre for Laboratory Animal Experimentation (SYRCLE)'s risk of bias tool can be employed for quality assessment of animal studies.

Another concern in evidence-based research is the external and internal validity of the articles. By asking, "Did the study have a proper research question?" you are considering the external validity, and by asking "Did the study answer the research question correctly and without bias?" internal validity is concerned. The most critical concern in the review of an article is the extent to which the results are acceptable. A study may have been conducted to the highest possible standards but is still biased. Some quality markers in a medical study, such as obtaining ethics committee approval, performing sample size calculations, etc., directly impact bias risk. Based on these concerns, Fresno's test was designed in evidence-based medicine. This test, also used in dentistry, has a total validity of 97% and asks questions for each stage of the research strategy. In total, 12 questions are designed in this test, and the time to answer the questions is 30 minutes.

The WWH (What-Who-How) scale is another tool for assessing the quality of the articles. It is designed with nine basic questions as a guide for reading the research section of the articles. These questions are directly related to the way research, design, and data analysis are performed. You can give each question a quantitative score and a short qualitative assessment.

1. What:

    1.1 What is the result of the study, and has this result been adequately tested and measured?
    1.2 How are the study findings presented, and do the findings respond to the claimed result?
    1.3 What are the clinical concepts and statistical concepts of the findings?

2. Who:

    2.1 What are the samples tested, and are these samples representative of the study population?
    2.2 Are the number of samples mentioned in the article reliable?
    2.3 Is the information provided in the article useful for patients, and does it have a clinical application?

3. How:

    3.1 How is the research question evaluated (experiment, observation, clinical trial)?
    3.2 Is the validity and reliability of the measurement tool introduced?
    3.3 How are the data analyzed? Were there more than two groups? Are the statistical tests used correctly?

Unfortunately, in many cases, researchers are forced to ignore the factors that affect the results of their research. Factors such as patients not attending on time, brushing irregularly, not eating the foods they were advised to eat during the study, misreporting the frequency of mouthwash consumption per day, smoking, and millions of other issues can significantly affect the results of the study. Still, the researcher has no choice but to trust the patient's words.

> The following is an example of the quality assessment used in a study about the "difference in failure rates between amalgam and composite resin posterior restorations" [17]:
> "The quality analysis of the included studies was conducted in accordance with the Newcastle-Ottawa scale (NOS), designed to be used in systematic reviews that include non-randomized, specifically cohort, studies. For the analysis, three main categories are addressed: selection, comparison, and results. For the selection and results categories, the studies may obtain one star/point for each item. For the comparison category, two stars/points may be assigned. According to NOS, the maximum score assigned to a study is nine stars/points (highest scientific evidence). Studies scoring six stars and above are regarded as high quality".

3.6   Extracting Data from Studies (Descriptive Analysis)

Data extraction in the systematic review means summarizing the information of evaluated articles using structured tables to give the reader a better gestalt for the evidence, facilitating the comparisons between them, and accelerating the next step, which is data synthesis. Furthermore, you can identify numerical data for meta-analyses. Moreover, by extracting the information of each article and filling the table, you may realize missing details in the existing evidence. This step must be carried out rigorously and systematically to provide a precise portrait of the studies.

The systematic review protocol guides the choice of data to be extracted. Based on your research question, you may need different data predetermined in your protocol. Since data extraction is a time-consuming step, it is important to limit yourself to the necessary data. Suppose you aim to "evaluate the osteogenic effect of a material on stem cells." In that case, you don't need to extract data related to cell proliferation, migration, inflammation, etc. You should only extract results of assays related to osteogenic differentiation like ALP activity, osteogenic markers' expression, or alizarin red staining. Therefore, you need some knowledge to understand what the author meant and identify the data that should be extracted. The data extracted from the articles produce the basis of the information utilized for data synthesis. Therefore, the main objective is to avoid making mistakes and to be able to spot them if they have been made during the data extraction process.

Data extraction should correctly display the information reported in the articles. Also, the extracted data must contain enough information so that each article is entirely understandable and can be analyzed. Two types of tables are required for this purpose:

- *Evidence tables:* Details of each article (including study design, PICO elements, results, and main conclusion) are provided in *evidence tables.* These tables are typically not included in your final review article. It is beneficial to prepare a data extraction form to consider all the information of the articles. The form systematizes the info collection and ensures a consistent procedure, regardless of who is responsible for extracting the data from each study. When several people are extracting data, the form should include instructions to standardize the work. These people should receive training on how to use the form. Likewise, the form must contain a space to enter the name of the person who extracts the data from each item. By filling out the form for two or three articles, you can understand its weaknesses. An electronic document is preferable, either an MS Excel file or software designed specifically for this purpose (e.g., RevMan: Review Manager). You may extract all the information related to the reference, the objective of the study mentioned by authors, and study design from each article and add your comments about them. By preparing a detailed evidence table, you can minimize the frequent return to the source articles. Examples of data to be extracted are study design, characteristics of the participants (number of participants, age, sex, etc.), location (geographic region, type of care setting), features of the intervention (nature, dose, frequency, duration, etc.), measurements (type, unit, time of measurement, etc.), results (unit, sample size, etc.), side effects (indicate if not reported), and other information like conflicts of interest, sources of funding, patient participation in the study, etc.
- *Summary tables*: Selected information from evidence tables will be provided in *summary tables.* These tables would describe the articles' results to answer your research question and will be included in the *result section* of the final review article. You should divide the research topic into subsections that will help you present results more precisely. Summary tables usually are prepared based on the PICO principles. Characteristics of populations, interventions, outcomes, or study design will be examined across studies. You can find common characteristics in the selected articles and cluster them along with those areas. Based on the similarities found in the included studies, they could be grouped in summary tables. The rationale for assemblage should be explained.

In summary, for extracting data from evaluated articles:

- Use your primary research question to realize which data need to be extracted.
- Data extraction requires sufficient knowledge of the subject.
- Predict the required sections of your summary tables.
- Use the PICOTS framework to select data: Population, Intervention, Comparator, Outcome, Timing, Study design.
- Describe accurately the process for extracting and summarizing data.

- Data extraction by at least two reviewers is ideal. Extracting data by one reviewer and checking by another one can also be a valuable method.
- In meta-analysis studies, a specific statistical strategy is required for data extraction.

> The following is an example of the data extraction used in a study about the "In vitro cytotoxicity of dental adhesives" [18]: "For data extraction, scientific and technical information items were tabulated and analyzed with Microsoft Office Excel 2013, and two reviewers conducted the analyses independently. The data extracted included the study year of publication and author(s), adhesives used, types of cell lines, cell culture methodologies, parameters evaluated for cell viability, use of controls, and study outcomes."

3.7   Data Synthesis and Evaluation

In the last step of systematic review, data from the included articles will be brought together to draw a conclusion. The findings of your review should be synthesized considering a recognized contribution to the issue. Descriptive summaries of studies related to a topic without meaningful analysis do not significantly contribute to the field. By synthesizing data, you can answer the primary question of your review research. The methods employed to synthesize the data should be previously indicated in the protocol.

Data synthesis compares, combines, and summarizes the results of included articles with respect to the research question and the PICO elements of included studies. The purpose of data synthesis in a systematic review is to express similar features among selected articles, explain the relationship between the obtained data, compare and combine them, translate the level of agreement in the data, look for patterns in data, produce the answer to the review question by integrating the data, and finally, assess the strength of evidence by evaluating the association of synthesis to the original review question. By considering the strength of evidence, investigating whether the observed effects are similar across studies, and examining the reasons for these differences, one can draw evidence-based conclusions. You should also consider reporting the limitations of the synthesis. Briefly, the systematic review report contains a "discussion" comprising the following elements:

- Summary of main findings
- The general interpretation of the results
- The strengths and limitations of systematic review
- Gaps in scientific evidence and recommendation for further research directions

You may perform a statistical synthesis via statistical methods, if possible, which results in a meta-analysis. The meta-analysis consists of grouping together the results of the various selected studies (concerning the same measure) to produce a single result called the "summative result." When studies are too heterogeneous (clinically, or methodologically) a meta-analysis cannot evaluate them.

The following is an example of the data synthesis used in a study about the "barriers and facilitators for repairs of defective restorations" [19]:
"Meta-analyses of the proportions were performed using Comprehensive Meta-Analysis 3.3.070 (Biostat, NJ, USA). Cochrane's Q and I^2-statistics were used to assess heterogeneity. Since heterogeneity was found high, random-effect models were used. To evaluate potential changes of the proportions through the years, meta-regression using the maximum-likelihood method was performed. Bonferroni correction was performed to adjust for alpha-inflation; as we performed three meta-regression analyses, p < 0.05/3, i.e., p < 0.017, was regarded as significant. Publication bias was evaluated using funnel plots as well as Egger's regression intercept test".

The conclusion of a systematic review should involve the implication of your evidence for practice or research. Figure 3.7 shows the flowchart of the systematic review process.

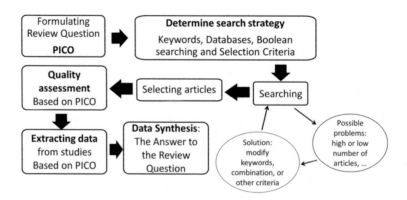

Fig. 3.7 The flowchart of the systematic review process

References

1. M. Torabinejad, N.A. Dinsbach, M. Turman, R. Handysides, K. Bahjri, S.N. White, Survival of intentionally replanted teeth and implant-supported single crowns: A systematic review. J. Endod. **41**(7), 992–998 (2015)
2. Levels of evidence—grading system *. J. Evid. Based Dent. Pract. **17**(2), A11–A13 (2017)
3. S. KC, X.Z. Wang, J.E. Gallagher, Diagnostic sensitivity and specificity of host-derived salivary biomarkers in periodontal disease amongst adults: Systematic review. J. Clin. Periodontol. **47**(3), 289–308 (2020)
4. Z. Lin, F.J. Strauss, N.P. Lang, A. Sculean, G.E. Salvi, A. Stähli, Efficacy of laser monotherapy or non-surgical mechanical instrumentation in the management of untreated periodontitis patients. A systematic review and meta-analysis. Clin. Oral Investig. **25**(2), 375–391 (2021)
5. J. Suvan, Y. Leira, F.M. Moreno Sancho, F. Graziani, J. Derks, C. Tomasi, Subgingival instrumentation for treatment of periodontitis. A systematic review. J. Clin. Periodontol. [Internet] **47**(S22), 155–175 (2020)
6. S.A. Nowrin, S. Jaafar, N. Ab Rahman, R. Basri, M.K. Alam, F. Shahid, Association between genetic polymorphisms and external apical root resorption: A systematic review and meta-analysis. Korean J. Orthod. **48**(6), 395 (2018)
7. S. Tumedei, D. Fabbro, Synthetic blocks for bone regeneration: A systematic review and meta-analysis. Int. J. Mol. Sci. **20**(17), 4221 (2019)
8. N. López-Valverde, J. Flores-Fraile, J.M. Ramírez, B. Macedo de Sousa, S. Herrero-Hernández, A. López-Valverde, Bioactive surfaces vs. conventional surfaces in titanium dental implants: A comparative systematic review. J. Clin. Med **9**(7), 2047 (2020)
9. M. Rezai Rad, S. Mohaghegh, F. Kouhestani, S. Reza Motamedian, Effect of mechanical forces on the behavior of dental stem cells: A scoping review of in-vitro studies. Mol. Cell. Biomech. **18**(2), 51–67 (2021)
10. T.T. Fröhlich, R.D.O. Rocha, G. Botton, Does previous application of silver diammine fluoride influence the bond strength of glass ionomer cement and adhesive systems to dentin? Systematic review and meta-analysis. Int. J. Paediatr. Dent. **30**(1), 85–95 (2020)
11. S. Kulkarni, M. Meer, R. George, The effect of photobiomodulation on human dental pulp–derived stem cells: Systematic review. Lasers Med. Sci **35**(9), 1889–1897 (2020)
12. A.R. Cocco, W.L. de Oliveira da Rosa, A.F. da Silva, R.G. Lund, E. Piva, A systematic review about antibacterial monomers used in dental adhesive systems: Current status and further prospects. Dent. Mater. **31**(11), 1345–1362 (2015)
13. F. Tabatabaei, Z. Aghamohammadi, L. Tayebi, In vitro and in vivo effects of concentrated growth factor on cells and tissues. J Biomed. Mater. Res. Part A. **108**(6), 1338–1350 (2020)
14. S.D. Currell, A. Liaw, P.D. Blackmore Grant, A. Esterman, A. Nimmo, Orthodontic mechanotherapies and their influence on external root resorption: A systematic review. Am. J. Orthod. Dentofac. Orthop. **155**(3), 313–329 (2019)
15. F. Tabatabaei, K. Moharamzadeh, L. Tayebi, Three-dimensional in vitro oral mucosa models of fungal and bacterial infections. Tissue Eng. Part B Rev. **26**(5), 443–460 (2020)
16. S.D. Ferreira, C.C. Martins, S.A. Amaral, T.R. Vieira, B.N. Albuquerque, L.O.M. Cota, et al., Periodontitis as a risk factor for peri-implantitis: Systematic review and meta-analysis of observational studies. J. Dent. **79**, 1–10 (2018)
17. V. Moraschini, C.K. Fai, R.M. Alto, G.O. dos Santos, Amalgam and resin composite longevity of posterior restorations: A systematic review and meta-analysis. J. Dent. **43**(9), 1043–1050 (2015)
18. I.P. Caldas, G.G. Alves, I.B. Barbosa, P. Scelza, F. de Noronha, M.Z. Scelza, In vitro cytotoxicity of dental adhesives: A systematic review. Dent. Mater. **35**(2), 195–205 (2019)
19. P. Kanzow, A. Wiegand, G. Göstemeyer, F. Schwendicke, Understanding the management and teaching of dental restoration repair: Systematic review and meta-analysis of surveys. J. Dent. **69**, 1–21 (2018)

Chapter 4
Writing a Research Proposal

4.1 What are the Requirements?

As we explained in Chap. 2, the second step of the design cycle is designing a plan or the research map by writing the research proposal. The requirements for organizing different proposal sections depend on funding agencies and academic institutions [1]. For example, in National Institutes of Health (NIH) grants, research strategy which consists of significance, innovation, and approach (specific aim: justification and feasibility, research design, expected outcomes, and potential problems) is limited to 12 (for R01, R15, and R34) or 6 (for R03 and R21) pages. However, each proposal typically answers several fundamental questions, which come in the following order:

– What do you want to do?—Title
– Why do you want to do this research? What is the foundation?—Statement of the problem, significance, rationale, relevance
– What is your goal? What answer do you expect to get? —Specific aims, hypotheses
– How do you do that? How are you going to collect data? And what are the advantages of your methodology?—Approach
– When do you expect to finish the project?—Timeframe
– Who will do what, and what resources do you need?—Budget, personnel, and environment

Since everything that needs to be accomplished will be announced in advance, and the study has not yet been completed, the research proposal is considered a project, and the grammatical time for writing the proposal is in the future.

© Springer Nature Switzerland AG 2022
F. Tabatabaei, L. Tayebi, *Research Methods in Dentistry*,
https://doi.org/10.1007/978-3-030-98028-3_4

4.2 Title

The proposal starts with the research topic or title. You want to get your idea funded, so the topic should be exciting and innovative and address a need (relevant) [2]. A good title should also be clear and concise.

1. Innovative
   Research should benefit the scientific and dental community. With this view, there is no benefit in research if it is not innovative. Consequently, when a researcher chooses a topic, he/she should check whether someone has already worked on it. Chap. 1, Sect. 1.3, can help you in this aspect.
2. Clarity
   The title should contain some of the facts addressed in the research project. For example, if the title includes words like "tissue engineering," the proposal's text should also talk about tissue engineering. Another critical point in choosing a topic is that we should avoid choosing emotional and interrogative titles. For example, choosing a title such as "Genetic Engineering in Dentistry: Disaster or Opportunity!" is inappropriate. The title should be precise and informative and allow the reviewer to understand the project overview. When concepts such as "material synthesis" are considered, the type of material and synthesis method must be determined in the title ("synthesis and characterization of a *3D-printed functionally graded* porous *titanium* alloy"). Therefore, a good research title can be decomposed objectively. It should also be understandable to the audience at all levels so that there is no ambiguity on the subject. Although abbreviations are not recommended in the proposal title, sometimes a catchy acronym could help the review committee recall the main components. You should avoid orientation or prejudice in the title. For example, suppose a researcher wants to study the osteogenic effects of a commercialized bone graft made by a recognized manufacturer. In that case, the title should be completely unbiased and not promote this material.
3. Concise
   The number of words allowed for the research title is usually between 5 and 15. In general, neither very brief nor very long titles are acceptable. Choosing a very short-term title like "tissue engineering" for research is ambiguous. However, long titles are more imprecise than short titles. Sub-title can be added to the main title (e.g., main title: Investigating the Differentiation of Endometrial Stem Cells to Odontoblast Cell; sub-title—in vitro study). By using the same words repeatedly in the body of your proposal, the reviewer will recall the keywords even after reading the document.

4.3   Statement of the Problem/Significance

The most critical part of a research proposal that impacts funding achievement is the problem statement, which is the first section of the proposal that should be written. Statement of the problem means presenting a problem that needs to be investigated, and justifying it so that the necessity of doing it can be felt. There is a question in the researcher's mind, which must be shared with the audience. This question may have arisen from the researcher's experience or was created by a supervisor or an institution. Defining the research problem is a prerequisite for clarifying and focusing on the topic chosen by the researcher. It is the answer to the questions: what research is to be done/why is this problem significant/why is this research relevant? The researcher must study the theoretical foundations of the problem and examine the research already done for answering these questions. Although it may take a lot of time and effort, if the researcher does not take this fundamental step seriously and fails to lay a good foundation for future activities, he/she will not be able to do the research properly.

The researcher must know for whom and what level and specialization he/she wants to explain the subject. The further the audience is from the researcher's field, the more time and content is needed to align the problem statement with the topic. Consequently, the audience is one of the primary tools for identifying a problem statement. While the audience of a dissertation proposal is completely expert, in grant proposals, the audience may be educated but not experts in a particular field.

Explaining the problem constitutes the central pillar of any research because the researcher forms all his/her research activities based on the problems he/she wants to solve. You should explain how your topic aligns with your previous studies and fits with your future career plans. A well-defined problem can stand out and be easily recognized by reviewers. This section should clarify the reader's mind about the background and history of the phenomenon in question, describe knowledge gaps that make research necessary, explain how the proposed research will improve knowledge, and prove that this research needs to be done. In other words, the researcher in this section must justify the implementation of the project in such a way that the budgeting organization is persuaded to provide financial resources. Therefore, the problem statement:

- Is an integral part of the selection of the research topic
- Is the essential basis for establishing the research proposal and will constitute a guide for the precise development of the research plan (research objectives and hypotheses, methodology, work plan and budget, etc.)
- Allows the researcher to describe the problem systematically, judge its importance and priority in the country or at the local level, as well as point out why the proposed research should be undertaken to solve the problem
- Facilitates the judgment of the proposal for organizations that contribute funding

The first sentence of the problem statement is crucial for catching the reviewer's attention. You can present the first phrase as a question and not a statement. For instance: "Why is periodontal regeneration challenging?"

The purpose of this section is to establish how the research question arises from current knowledge on the subject. After introducing the problem, you should indicate where it comes from and how you chose it among so many others in this field of research. You should then evoke the motivations which aroused your interest in the topic. The progression of the ideas that gave birth to it must be presented in a logical sequence and a concise manner. You must also show that your topic is original and novel. The literature review you have done before starting the design cycle would help you write this part of the problem statement by leading you to the description of the problem and showing that you can add to the knowledge in this field by performing your research. When you have a good awareness of existing literature, you can demonstrate the originality of your topic and explain the importance of the subject, the dimensions of the problem, and its characteristics. You can also describe any results you have already obtained in the research area of the proposed study. By addressing the challenges, you can clarify the value of your solution to overcome these challenges.

The researcher expresses the relevance or scientific scope of the topic by indicating how this topic falls within the scientific concerns of other researchers, how this research will contribute to the advancement of knowledge, or how it provides answers to the concerns of social decision-makers, politicians, practitioners, etc.

How should the problem statement be written in the research proposal? Figure 4.1 shows the answer to this question. The logical order of writing the problem statement is:

1. Defining the problem clearly in one paragraph (What is the problem? What question do you hope to answer?).
2. Prioritizing and justifying the problem. Mention the impact/importance of the problem, its current interest, distribution and severity, and socioeconomic effects (Is the research question important? Why is it essential to fix it?).
3. Showing that you reviewed the literature and are knowledgeable about the available studies in your field. Have other researchers addressed the same or a similar problem? Is there currently any attempt to fix the problem? If so, what are the outcomes and shortcomings (the side effects or consequences of the current solution)? What is the discrepancy between the existing and expected outcomes (knowledge gap)?
4. Formulating the rationale of your study. What is your potential solution? What is your assumption about the answer you expect to get? What are your long-term

Fig. 4.1 The summarized process of the problem statement

goal, overall objective, and central hypothesis for achieving the objective? How will this research bridge the gap? What is your experience in this area? Do you have any compelling pilot data that form the basis of your hypothesis?

5. Showing the values of your solution and its contribution to the body of knowledge. To what extent does your research combine with current expertise in the field of research concerned? What contribution will your research make in the science, clinic, or industry? What is the innovation (extending knowledge, improving an instrument or a treatment)? What is the next research step? Lack of contribution means that study's objectives should be changed.

Through the above steps, the researcher can convey that his/her question is significant and should be investigated.

Let us suppose that you phrased your research question as follows: "In periodontium defects which include gingiva, alveolar bone, cementum, and periodontal ligament, does an integrated four-layer model have a different effect in comparison to guided bone regeneration (GBR)?" The first sentence should be catchy and attractive for the reviewer. You can state the health impact of complete periodontium regeneration in periodontal defects as a critical problem to solve. Secondly, you should explain that a periodontal defect may be a complex multi-tissue loss of combined soft (gingival and periodontal ligament (PDL)) and hard (bone and cementum) tissues caused by disease. Then you should determine the importance of this problem by mentioning its prevalence in society and its economic and social effects. Now it is time to refer to the actual treatments like GBR/GTR. Citing landmark studies and presenting the contribution of other researchers to fix the problem is important in this step. Afterwards, you can explain the current solution's drawbacks, critical needs, or existing gaps in knowledge, such as the lack of complete compartmentalization between the different parts of the multi-tissue defects during the treatment because each of the four tissues requires its own cell accommodation and specific environment to be grown, which cannot be provided in current treatment methods. Phrases like "these studies were limited by…," or "a gap remains…" can be used in this part. Next, you can address the hypothesis, long-term goal, and objective. You hypothesize that an integrated four-layer construct is more effective than GBR. A hypothesis translates the statement of the problem into an accurate and unambiguous prediction of the expected results. Your preliminary data could support your hypothesis. Your long-term goal could be improving periodontal regeneration. Remember that testing the hypothesis could not be the objective. The objective should address the gap by highlighting the research product. Your objective could be "complete compartmentalization between the different parts of a model for multi-tissue defects." Finally, you will address your study's "so what" or rationale by highlighting your research's contribution to society or the clinic.

Figure 4.2 shows the relationship between the preliminary data, long-term goal, and objective. However, you don't need preliminary data for all funding applications. While NIH R01 requires demonstration of researcher expertise by describing previous studies, exploratory grants like R03 and R21 can be used for pilot studies,

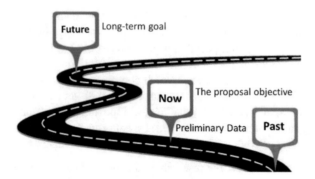

Fig. 4.2 The relationship between preliminary data, long-term goal, and the objective

new tools and treatments, or the feasibility of a protocol. Data gathered using these sources can be employed as preliminary data for an R01 application [3].

In NIH grants, "significance" is usually more focused on knowledge gaps and how these issues will be tackled by completing specific aims. The purpose of this section is to explain the rationale for the research in terms of the importance of the expected outcomes and to show that you have something unique to offer. Addressing the relevance of the proposed research to human health/disease and the objectives of the funding agency should be considered in the "significance" section. Read the funding announcement carefully and use their own terminology and language to address your project's alignment with their mission [4, 5].

4.4   Specific Aims

Usually, the research question is too broad, and it is necessary to refine the question to identify the overall objective and specific aims for the proposed research. A distinction must be made between the *general objective* and *specific aims*. Defining the objective of a project is a particularly strategic phase. It constitutes the starting point from which the experiment can be structured, and it also serves as a basis for the evaluation phase (the last step of the design cycle). The research objective determines the effect of changes in one or more independent variables on one or more dependent variables. It is "what will be accomplished by the research project" and relates to the questions that the researcher wishes to answer through the proposed study. On the other hand, the identified gaps in the knowledge or critical needs should be addressed by specific aims which fulfill the objective. The specific aims specify practical steps the research team will need to complete to achieve the overall objective.

Expressing specific aims means breaking down the overall research objective according to their primary areas of intervention to more specific and detailed aims which will be reached in the research process. The researcher must acquire a set of aims to achieve the overall objective. If the research aims are well articulated, the researcher will find out "what kind of information he or she needs" and "how to look for it," which will pave the way for research.

Two critical factors should be considered in writing the specific aims:

1. First, each aim should not consider only one outcome; otherwise, its failure means a dead-end aim. For instance, if you aim to characterize "the quality of the interface of methacrylamide adhesive materials synthesized with aldehyde functionalities," you only consider one outcome. You can modify this aim as follows: "characterize the quality of the interface, as well as collagen crosslinking and proteolytic activity of methacrylamide adhesive materials synthesized with aldehyde functionalities" (https://grantome.com/grant/NIH/K02-DE025280-05).
2. Second, failure in one aim should not negatively affect other aims or prevent the completion of the project. For this purpose, the specific aims must be independent of each other even though they are related. In a proposal about a new triple antibiotic paste for endodontic treatments in pediatric patients, if one aim is to develop this new paste and the second aim uses the paste for endodontic treatment, failing in paste development results in the subsequent aim's fail. You can avoid this by providing your preliminary data and showing that you seek to refine further or develop an already successful approach. You can also highlight the gaps in prevalence, effectiveness, and safety of current treatments and address these three gaps in specific aims:

   – Does triple antibiotic paste reduce endodontic treatments failures or increase the time before failure?
   – Does triple antibiotic paste affect *E. faecalis*, or it is effective against other microorganisms?
   – Is triple antibiotic paste safe for periodontal tissues, or it is cytotoxic for bone marrow stem cells?

By following this approach, each aim will consider two outcomes, and aims are related but not dependent on each other. In another example, if your objective is to "improve the characteristics of dental pulp cells by using low-power lasers," specific aims will include the following:

– Measuring the effect of laser on cell proliferation or cell migration
– Investigating the effect of laser on osteoblastic differentiation or angiogenesis

As we mentioned in the above examples, it is better to use action words (like measure, investigate, examine, evaluate, assess, develop) instead of descriptive verbs (such as describe, define) for stating the specific aims. Example:

– Well-designed aim: We will compare the role of genetic polymorphisms in the interleukin (IL6) in type 2 diabetes mellitus and periodontitis.
– Unacceptable aim (descriptive, one outcome): We will determine if genetic polymorphisms in the IL6 play a critical role in developing periodontitis.

– Unacceptable aim ("fishing expedition" and descriptive): We will identify IL6 gene polymorphisms in biopsy tissues obtained from a cohort of 20 patients with periodontitis.

An associated hypothesis, a summary of the approach, expected outcomes, and predicted challenges could be included in each aim. The proposed method must test the hypothesis, which means it must be possible to identify which parameters to estimate, its standard deviation, and the sampling distribution. Therefore, by detailing the overall objective to specific aims, the researcher also talks about the feasibility of conducting research. The researcher should mention the effect size in the expected outcome. The effect size means the minimum clinically important or reasonable differences between the groups that the researcher expects to notice. This effect size should be determined with scientific knowledge and be based on the previous publications on a related topic or be produced by the researcher in a pilot study. Overall, by studying each aim (3–5 sentences), the reader should find out what the researcher wants to do, how he/she will do it, and what he/she expects to find.

Usually, pilot studies and short duration proposals with a lower budget (e.g., R03) have two specific aims, while larger grants like R01 need three to four specific aims. Adding a figure which shows the relationships between variables can provide visual support for specific aims. Talking about your aims with colleagues, scientific review officers, and program officers can help you in their development.

In summary, follow these rules in writing of specific aims:

- Specific aims must be related but independent of one another.
- Specific aims should be measurable, which means having a predictable endpoint.
- Specific aims should be realistic (not fishing), which means avoiding limitless (broad) variables.
- Each aim should convey *why* this part of the research is required.
- Specific aims should not provide the details of methods or techniques.
- Consider different potential outcomes for each aim.
- Do not use descriptive aims like "to describe..." or open-ended aims like "to study...".
- Articulate a broad aim followed by a focused hypothesis for each aim.
- Show that the hypothesis is plausible based on your preliminary data.
- Show that you considered unexpected troubles by proposing alternative approaches for each aim.
- Use bullets to outline each specific aim.
- Add a figure.

Specific aims of NIH grants should be presented on one page with a "specific aims" title. This page may include four paragraphs named as introductory, rationale, specific aims, and overall impact paragraph [5–7]. For each paragraph, we incorporate an example from a granted proposal (https://www.niaid.nih.gov/sites/default/files/R01-Faubion-app.pdf):

1. *Introductory Paragraph:* The first paragraph consists of the research topic and its importance (*Hook*), the background (*What is known*), the knowledge gap (*What is unknown*), and the importance of filling this gap (*Critical need*). You should explain what has been done until now, why it is not enough, and how the problem is related to the funding agency mission.

Example

Hook: "The transcription factor FOXP3 is critical for regulating numerous debilitating human immune-mediated diseases, the prevalence of which together affects over 8.5 million people (1 in 31 US residents)." *What is known*: "In inflammatory bowel disease (IBD), chronic intestinal inflammation indicates aberrant in vivo FOXP3+ T regulatory (Treg) cell function (1). Similarly, pro-inflammatory signals in vitro impair Treg function (2). Our lab was the first to characterize the essential role of the histone methyltransferase (HMT) EZH2 in the epigenetic regulation of FOXP3 (3). Recently published work extended our observations indicating a pivotal role for EZH2 in FOXP3 repressor function (4)." *What is unknown*: "however, the regulation and biological impact of the FOXP3-EZH2 pathway to IBD is unknown." *Critical need*: "This knowledge is essential given the apparent loss of function of Treg cells in inflammation."

---

2. *Rationale Paragraph:* The second paragraph includes *long-term goal* (shows your vision for the future: the problem is big enough to occupy you for decades), short-term *objective* (link back to the knowledge gap), your preliminary data (evidence for hypothesis), *central hypothesis* (link to the objective and reflect the area of your expertise: formulate the central hypothesis in response to your research question and based on your own preliminary data or existing literature), and the *rationale* (what does conduct of this research offer? Once your project has been done, will you be able to advance your field based on the knowledge gap that you have discussed?).

Example

"Our *long-term goal* is to dissect epigenetic mechanisms regulating Treg cellular differentiation and function, particularly within the setting of GI inflammatory diseases, as these discoveries will facilitate the design of human cell therapy trials for IBD. Consequently, the *objective of this grant* is to characterize the role of the epigenetic regulator EZH2 in Treg suppressive function. These investigations are strongly supported by *preliminary data* demonstrating that: (1) EZH2 is required for Treg suppressive function, (2) IL6 signaling leads to phosphorylation and inhibition of EZH2, (3) lymphocytes isolated from the intestine of IBD patients demonstrate activation of IL6-induced gene networks and loss of EZH2 HMT function, and (4) conditional knockout of EZH2 in FOXP3+ T cells leads to in vivo immune dysfunction. Based on these compelling data, we propose the *central hypothesis* that EZH2 plays a critical role in Treg cells' homeostasis, and the disruption of EZH2 function by inflammatory signaling pathways contributes to IBD. Our *rationale* is that identification of the mechanism(s) to restore Treg suppressive function in the setting of intestinal inflammation will offer new therapeutic opportunities within the field of IBD."

3. *Specific Aims Paragraph:* In the third paragraph, you should mention your specific aims (to evaluate…). Each specific aim should include a hypothesis (our hypothesis is that…), the rationale (this hypothesis is based on …), experimental approach (to test this hypothesis we will…), and expected outcome or impact (the expected outcome of this aim is … and it will lead to…). Each aim should demonstrate its concordance with the central hypothesis by explaining why it is proposed, what it will produce, and how it will answer the hypothesis. You should not emphasize the process of doing the aim. You will determine the "effect size" or the "meaningful difference" from the minimum anticipated difference between the groups in the expected outcome. This assumption should be clinically or scientifically relevant to the clinical scenario or an existing standard as closely as possible.

Example
This aim *hypothesizes* that EZH2 is recruited to target genes in complex with FOXP3 (Fig. 2). Our *approach* will be domain mutational analyses of EZH2 and FOXP3. We will test the function of these mutations by both molecular (target gene repression) and cellular (Treg suppressor function) assays. The *rationale* for this aim is that understanding the mechanism of FOXP3 repressor function will lead to a therapy to enhance Treg cell function in vivo. Our *expectation* is to identify the protein domains within EZH2 and FOXP3 required to repress target genes and enhance Treg function.

4. *Overall Impact Paragraph:* The fourth paragraph contains the expected outcome (filling the gap), innovation, or the impact of your proposal on the advancement of the field. You can highlight how your proposal would address the funding agency objectives or strategic initiative. Depending on the mechanism and type of your proposal, you may also emphasize the grant's effect on your career development and exhibit a larger research agenda based on the proposed study. It is better to write the expected

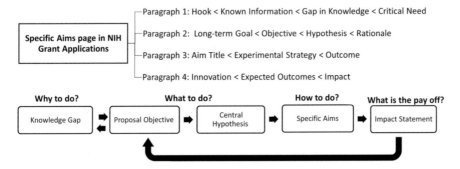

Fig. 4.3 Four paragraphs of specific aims page in an NIH grant application and the link between different parts of this section

outcome in the direct present tense, while the impact, which is linked to the critical needs and proposal objective, could be written in the future tense.

> **Example**
> Upon conclusion, we *expect* to understand the role of EZH2 in Treg loss of function in the setting of active inflammation. This *discovery will* stimulate new areas for experimental therapeutics in human chronic inflammatory diseases.

The components of each paragraph and the link between different parts of the "specific aim" section in an NIH grant can be summarized as shown in Fig. 4.3.

4.5   Literature Review (Justification and Feasibility)

The "Justification and Feasibility" section is a combination of a review of relevant literature and your own preliminary data. You should consider the difference between this section and the statement of the problem. In the problem statement, the foundations of the subject are examined, the background of that problem and the knowledge gap are determined, and finally, the researcher's hypothesis, how it addresses the gap, and his/her questions are discussed. However, in the literature review section, you will focus on the published works related to your aims to show your up-to-date awareness of the relevant literature and justify the aims. Here, you should be able to summarize your own studies related to the subject of research and interpret the work of others in a storytelling approach that displays the work done so far. This section will convince the reviewer about the solid foundation of your research question in existing knowledge, why you are doing this research, and how your work will impact the literature.

Fig. 4.4 Funnel shape of
literature review structure:
moving from the broad
issue to how your research
addresses the gaps

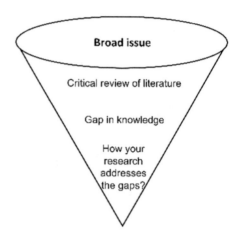

Reviewing the literature in the first steps of research (Chap. 1, Sect. 1.3) gives the
researcher confidence about the novelty of his/her topic. No matter how innovative
our theory may be, it does not mean that others have not stated something about it.
There is always a source for our topic, but it may not be complete, which is natural
for an innovative subject.

For structuring this section, you can shape it as a funnel, moving from more
general resources to more focused studies (Fig. 4.4). The first paragraph of this section could highlight the broad issue which can grab the reviewer's attention. You can
then summarize the current articles, which have a logical and reasonable relation
with the title of your research. Compare the approaches applied for analyzing the
research question, demonstrate the disagreements reported in the articles, and critically evaluate them by mentioning their strengths and weaknesses in terms of per-
suasiveness, validity, reliability, or appropriateness. Do not think descriptively by
listing authors and their work but think analytically and critically by offering your
voice to showcase the field. Next, you can review your own studies and determine
their relation to the literature. Your preliminary data, which can be published or
unpublished, demonstrate your area of expertise in scope and support your central
hypothesis. Finally, you can conclude by showing how your research addresses the
gaps you have recognized and demonstrating the originality and novelty of your
proposed research.

Note: There is no need to list all the articles related to your research. If there are
100 sources related to your project, it is enough to mention one-tenth of these
sources in the literature review section.

4.6   Research Design

Various activities included in the research process should be put in a complete, yet flexible framework named "research design." By choosing an appropriate research design, you can answer your research question. In other words, "what you will do to answer the research question" forms your research design. At this step, you should know which type of study (Sect. 1.1) is most likely to answer your question and give the highest quality results. The rationale for experiments should be explained in research design, and control groups for all experiments must be defined. The study subjects or participants, sampling method and calculation of sample size, methods to conduct the research, data collection techniques, and data analysis method are discussed in this section. Also, a discussion of potential pitfalls and alternative approaches should be included in this section. Research design in NIH grants should be tied to the specific aims. At the end of this section, the reviewer should be able to answer the following questions:

- Is the approach—as described in the research design—able to answer the question asked?
- Can the specific aims be achieved within the framework of the designed approach and with the methodology proposed in the research design?
- Can the study be carried out with the resources identified?
- Once the study is completed and the results analyzed, will the researcher have sufficient statistical power to validate the generated data?
- Will the outcomes of the research, directly or indirectly, have a favorable impact on the field of study?

It is helpful to keep the above questions in mind throughout preparing this section by putting yourself in the reviewer's place. While subjective deviation from the research design is not advised, an objective deviation of the research design during research is possible.

4.6.1   Type of Research

The choice of research type is at the heart of the research plan, and it is probably the most crucial decision a researcher must make. Here, you must specify the type of study selected for the research (descriptive, analytical, experimental, or a combination of them) that we explained earlier in Sect. 1.1. Each of these research types would require a different research design [8]. For example, to test a hypothesis, the researcher can assign the independent variable, or exposure variable, to several subjects in the study and omit it for other subjects (controls) while controlling the confounding variables. This strategy constitutes an experiment in which the hypothesis is tested by *intervention*. Another researcher may choose to compare individuals exposed to a risk factor to those not exposed when analyzing the incidence of disease in these groups. In this case, it is an *observational analytical study* to see if the disease is related to exposure. This type of study also includes hypothesis testing. Yet another researcher might simply describe the distribution of a phenomenon or the result of a program. It is then a *descriptive study* without intervention or prior hypothesis.

Remember that the type of study should be chosen according to the objective of the research. For determining the most appropriate study design, your research question should be well-formulated (Sect. 2.2.2). A well-formulated research question will help you specify the objective and decide if you want to perform an experiment (experimental) or observe without intervening (observational). You should be able to explain why this type of study was chosen over other possible ones.

4.6.2 Subjects, Sampling Method, Sample Size

This section should provide [9]:

- A detailed explanation of the study population and definitions of inclusion and exclusion criteria for participation
- A description of the method of sampling and assignment
- A realistic estimate of the sample size

Here, you should explain what subjects you will study: stem cells, oral bacteria, samples of dental materials, extracted teeth, animals, patients referred to the dental clinic, or elementary school children? The population from which the study participants are drawn should be carefully selected according to the research objectives. The success of the investigation will depend very much on identifying this population. You need to determine the exact characteristics of the study population by defining the inclusion (required characteristics) and exclusion (unsuitable characteristics) criteria. These criteria are features, variables, or external factors that align with your study's aims or interfere with them and define the eligibility or ineligibility of samples to participate in your research. Being specific in writing these criteria is very important. Suppose that your study population is patients with gingivitis. For inclusion criteria, consider explaining that "subjects will be included in the study if they have gingivitis diagnosed by a dentist and have had at least 15% of marginal bleeding in the six surfaces of teeth." To guarantee that your criteria are reasonable, you can refer to literature with similar goals and provide a rationale for the defined criteria.

As the population of interest is usually vast and it is impossible to work on all of them directly, most research studies involve observing a *sample* from a defined population. Once the inclusion and exclusion criteria have been defined, you need to select samples from that population to enter the study by *sampling*. The samples are a limited member or subset of our study population chosen based on their representativeness and accessibility. The study's conclusions are based on generalizing the results observed in the sample to the total population. Consequently, the quality of the constitution of the sample and, more significantly, its representativeness would define the accuracy of the conclusions. If you choose samples carefully using a correct procedure, you can generalize the results to the whole population. Accordingly, the first concern in selecting an appropriate sample is that this sample is representative of the population. Each variable considered must have the same distribution in the sample as the population from which it is extracted. It is, therefore, necessary to know the variables and their distribution in the population. The statisticians have proposed means to give a reasonable guarantee of representativeness and determine how we should choose the individuals constituting the study sample (sampling method) and what size the selected sample should be.

How to choose the sample from the population? The *sampling method* can be probable (random) or non-probable (judgmental). Random sampling increases the validity of research findings so that results can be generalized to a larger population. When a known chance (greater than zero) of participation has been considered for every member of the population, your method is probability sampling, which includes:

1. *Simple Random Sampling*: This is the most common and simplest sampling method. In this method, subjects are selected from a homogenous population with an equal probability of selection. Software has recently been developed for taking random samples from a given population. The simple random sample has the advantage of being easy to apply, being representative of the population over time, and facilitating direct analysis of the data without any intermediate steps. The downside is that the sample chosen may not faithfully represent the population, especially if the sample size is small.

2. *Complex Random*: Could be *systematic* (determining the sampling interval of a homogenous population), *multi-stage* (when your study involves various methods of sampling in different groups, combining two or more probability techniques, which can be done in several stages), *cluster* (when the study concerns large populations having a large geographical dispersion, clusters can be identified and the study will include random samples of clusters, so each member of the cluster will take part in the study), or *stratified* (when the study population is heterogeneous, its size is considerable, and we have some information about the distribution of a particular variable (e.g., gender: 50% male, 50% female), it may be advantageous to choose simple samples randomly in each of the subgroups defined by this variable. By choosing half of the sample from men and a half from women, we ensure that the sample is representative of the population concerning gender).

In case you are selecting non-randomly the samples and not giving a chance to all members of the population for participating in the study, your method is *non-probability sampling* which includes: (1) *volunteer sampling* (selection of samples among those who volunteer themselves), (2) *convenient* or accessible sampling (sampling of those most accessible), (3) *snowball* sampling (every participant is referred to the researcher by the previous one), and (4) *purposive* sampling (selection of the sample based on researcher judgment) which could be *quota* (among the groups in the society based on specific criteria, for example, among men and women) or *matched sampling* (to take a control group).

It is essential to be aware of the difference between *random sampling* and *random assignment/allocation*. As mentioned before, the process of selecting people into the study (samples from the population) could be random sampling. After selecting subjects, you should also determine the method for their random assignment to one of the study/control groups. In scientific research, using control groups improves the validity of conclusions. Control groups include comparable units from the same population but differ in some respects, such as exposure to risk factors, use of therapeutic measures, or participation in an intervention program. Some descriptive studies (studies of existing data, surveys) are without control groups. However, control groups are necessary for all analytical studies, experimental studies of drug trials, research on the effects of intervention programs and disease control measures, as well as in many other investigations. In an experimental study, the control group includes subjects not receiving the intervention but who resemble the experimental group in all other aspects. The subjects for the experimental and control groups should be chosen and assigned at random to each group. In the split-mouth randomized trial, the unit of randomization is the site in the oral cavity. Comparison made within patient in this type of study can eliminate the inter-subject variability and increase the power of the study.

It is necessary to estimate the required *sample size* to answer the question. The sample should be of sufficient and optimum size to yield meaningful results and permit the application of statistically significant tests. *Sample size (n)*, the smallest number of samples required for testing the hypothesis, is based on assumptions. Statistical formulas used to calculate sample size vary based on the type of variable (quantitative or qualitative), the type of study (descriptive or analytical), and the specific objectives. Justifying the estimated sample size is very important. If the number of samples is too large, you may see a falsely significant difference between groups when there is no difference. This *type I error (alpha error)* results in the incorrect rejection of a null hypothesis (while true) and could have a harmful impact on science. Also, if the number of samples is insufficient, the result would not show a significant difference between groups while there is one. This *type II error (beta error)* results in the false acceptance of a null hypothesis with a less detrimental impact than the false-positive error. The *statistical power* determines the probability of a true rejection of the null hypothesis when it is false. Generally, we aim to minimize both α and β; however, these errors work in reverse. If one decreases, the other tends to increase. Avoiding type I error is more important than type II. Therefore, the investigator plans to have the desired level of α and minimize β for that condition. The choice of α and β is made after determining the consequences of these errors while it should be fixed before starting the study. Usually, a value of 0.05 for *p*-value or α error (= accepting the probability of a false-positive result) and a value of 0.20 for β error (= accepting the probability of a false-negative result), which corresponds to a power of 80% are acceptable for calculation of sample size. Designating the expected outcome (explained in previous steps) is also helpful for calculating the sample size. If your expected outcome is to reduce infection rate, the

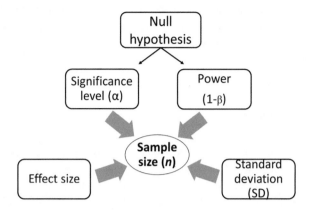

Fig. 4.5 Factors required for determining sample size

required sample size is much larger than if you have a score (e.g., pain score). We can determine the effect size (clinically relevant assumption) from the minimum anticipated difference between the groups mentioned in the expected outcome. A higher expected difference would result in a smaller number of required sample sizes. Once you have fixed the alpha, power (1-β), and effect size, you can use different formulae and software for estimating the sample size. There are also some free tools for calculating the sample size (e.g., http://www.stat.uiowa.edu/~rlenth/Power/ and http://www.r-project.org). Usually, including extra members in the sample size would account for samples that may be dropped out of the study. Overall, to determine the sample size (Fig. 4.5), you need to:

- Specify the effect size in the expected outcome: there is an inverse relationship between the effect size and the sample size.
- Specify the experimental material's standard deviation (SD) or variability based on the previous studies: a direct relationship between SD and the sample size.
- Specify the power (80–90%): a direct relationship between the power and the sample size.
- Specify the significance level (1% or 5%): an inverse relationship between the *p*-value and the sample size.

Collaboration with a statistician is essential, especially in clinical research, because it is more challenging to analyze the results in this research. Therefore, the required sample size should be determined with the cooperation of a statistician before starting the study. Since some tests are performed in a certain number as standard, it is not always necessary to decide on the sample size via statistical methods. For example, in the MTT test (which is a standard test to evaluate the toxicity of dental materials), according to the standard (ISO 10993-5), the sample size is always $n = 6$. Also, in microbial experiments, such as disk diffusion tests (based on CLSI standard), the sample size is three, and the test will be repeated two times. The average area of inhibition of growth around the disk will be reported. It is prudent to increase the number of samples considering the risk of sample/subject loss (presence of voids or crack in material preparation, sample contamination in a cellular/microbiological study, unexpected death of animals, patients dropping out of the study in clinical trials).

4.6.3 Data Collection

Depending on your type of research, which could be exploratory, descriptive, or analytical, the data collection method should be chosen from the beginning of the study. It is essential to establish how the data will be collected.

Instruments are the tools by which data is collected. These are, among others: questionnaires, medical exams, different types of laboratories experiments (microbial, cellular, material properties, etc.), animal studies, screening procedures, bioassays, clinical trials, or manufacturing processes. This complexity makes the result of each test highly dependent on the design of the test. You are conducting experiments to find the relationship between variables. Therefore, different types of variables must be considered and controlled in the experiments. For example, the type of cells used to measure the toxicity of metal ions released from metal materials or the ratio of the number of cells to the amount of the substance can significantly impact the result.

The standard instructions (ADA, ISO) and guidelines are available to evaluate the specific materials or conduct animal studies or clinical trials. Each instruction controls some of the characteristics. These standards describe how mechanical, physical, electrochemical, and biocompatibility tests are performed using laboratory, animal, and human methods. To test your hypothesis, you may have found several alternative experimental methods in the literature. You must choose the most straightforward one according to your abilities and the available facilities. This section should clearly describe what procedure will be performed on the samples/subjects and how the samples or the effect of treatment will be evaluated. If you use a standard method already described or follow the manufacturer's instruction, references must be provided.

Usually, any researcher who wants to market or register a new substance, biomaterial, or treatment must complete a series of tests to be allowed to enter the market. These tests are primarily laboratory; in the second stage, tests should be performed on animals, then the materials are tested in the animal's mouth with precisely the same clinical use. Finally, experiments will be performed on volunteer groups of humans. Before entering the first phase in humans, considerable research should be done in vitro and in animals to establish that the new material is effective and potentially beneficial for humans. Full compliance of the laboratory tests with standardized test conditions, maximum simulation of laboratory environments to the oral cavity's conditions, and a complete description of the test method can increase the validity of such research so that the results can be generalized to the clinical situation. Because laboratory tests are performed outside the living body, many of the complex interactions that affect biological responses in the body are not present. Animal tests allow for many complex interactions between the biological environment and biomaterial. But it is difficult to control the variables in these tests. Animal studies and cohort are beneficial, but the best experiments and research are clinical trial studies that provide the most acceptable answers but are also very costly and time-consuming.

The researcher should be aware of the *variables* and know how he/she wants to measure them. Variables are the characteristics of the study population (subjects) that we propose to study. For the success of the research study, it is necessary to have a clear and precise definition of variables, the method of measuring them (simple observation, laboratory measurement), and the unit of measurement. An *independent variable* is a variable that is monitored, exposed, intervened, treated, or considered as a risk factor (cause) to see what effect a change in this variable will have on variables postulated as being dependent. A *dependent variable* is a variable whose changes depend on the independent variable. It shows the outcome of monitoring, response to treatment, or disease after exposure (it is expected to change because of the intervention and is not under the researcher's control). A logical relationship between independent and dependent variables is required (such as the relationship between a person's diet and tooth decay). The researcher should also be aware of the interfering (confusing) variable that could influence or confuse the effect of the independent variable on the dependent variable and how he/she wants to control its impact. A confounding variable is an independent variable (other than the assumed causal variable in the hypothesis), which can affect the dependent variable, but whose distribution is systematically correlated with the causal variable in the hypothesis. For example, your dependent and independent variables are diabetes and periodontal disease, but you realize that sex, age, and smoking affect both variables. To deal with these interfering variables, you can either eliminate, match, or examine them in all groups. For elimination, any sample in the control and experimental groups with the intervening variable (smokers) is excluded from the study. For matching, only samples with the intervening variable are selected in both groups (females to equalize the effect of sex). You can also examine the two groups in terms of the interfering variable (age). Note that it is impossible to consider all variables other than those suspected to be causal variables in most studies. Through randomization, many of these confounding variables will be evenly distributed between the exposed and control groups. Procedures taken to avoid bias should also be conveyed elaborately.

Another essential point in data collection is the *validity* and *reliability* of the tools designated to measure variables (are we measuring the suitable characteristic? Accurately?).

Validity means that the tool of measurement can precisely measure the desired variable. If the measurement device does not have sufficient validity for the selected feature, the research results will be invalid. For example, a digital scale used to measure the weight of objects in milligrams is valid when it shows the exact weight of the object. Therefore, if we use a not zeroed scale, the weights measured with that scale will be invalid (biased).

Reliability indicates the consistency of the results of two tests repeated in the same conditions. If repeated measurements of a characteristic under identical conditions give similar results, the measurement can be reliable. Your invalid digital scale may be reliable if it shows the same number every time you measure a specific object. If we repeat independent observations and determine the probability distribution, the standard deviation of the observations gives a measure of reliability.

The validity and reliability of tools may be ensured by a specific description of inclusion and exclusion criteria, calibrating the devices, adopting standard methods and proper sampling, selecting appropriate time scale for the study, and carefully following the manufacturer's instructions when using the test kits. Researchers also use a variety of methods to ensure the validity and reliability of the tools:

A. Using Test-Retest and Parallel Forms

In the test-retest method, the researcher can select one group of samples or two groups of homogeneous samples at the same time and evaluate them by a tool in two different time points. For example, after isolating dental pulp stem cells from a tooth, you can assess the reliability of your PrestoBlue kit on a defined number of these cells after 24 h seeding. After 1 or 2 weeks, you can retest the PrestoBlue kit using the same cells with the same number and after 24 h.

In the parallel forms, the same phenomena in the same sample group may be assessed using more than one assessment method. For example, you can evaluate the cell proliferation of dental pulp stem cells isolated from the same tooth at 24 h using PrestoBlue, MTT, and Alamar Blue assay. Then, the results can be compared with each other to calculate their correlation coefficient. If the results are similar and the correlation coefficient is high (more than 90%), we can say that the tool has validity and reliability.

B. Using the Method of Comparison with Control

In this method, the researcher needs a positive or negative control group to measure the validity and reliability of the tool. The control groups will increase the credibility of your work. A positive control is a well-known substance or method that provides expected results. For example, a new dental cement's bonding strength and solubility can be compared to zinc phosphate. The choice of control group depends on your objective. If you have synthesized a new biomaterial and your goal is to study its cytotoxicity, positive control is a substance that has a cytotoxic effect (distilled water) on cells. On the other hand, if you want to show the positive impact of the synthesized material on cell proliferation, cells in a standard culture medium are the positive control group. A negative control is usually a substance or method in which no reaction or harmful effects should occur. For example, the negative control in the disk diffusion test of an antibacterial extract could be the sterile paper disk. Reaction in the negative control indicates an error in the test process. In some cases, you may have to compare the test item with a gold standard while considering the positive and negative controls. For example, in examining the cellular toxicity of a new pulp capping material, in addition to positive (the group of cells treated with distilled water as a toxic substance) and negative (cells fixed with glutaraldehyde) controls, you may consider the standard culture medium as a gold standard.

C. Using the Pre-test Method (Pilot)

Through this method, various issues of the research design are examined preliminarily, one of which is the validity and reliability of the measurement tool. The pilot study results can also help you estimate the sample size or modify the data collection method.

In summary, the data collection section includes:

- The precise definition of all variables and methods
- A pilot study to test the methodology and the instruments used
- A discussion of the validity and reliability of the proposed tools

4.6.4 Data Analysis

By conducting experiments, you will have access to primary/raw data. Data are values taken by a variable (compression strength, contact angle, porosity, roughness, etc.). After collecting the raw data, we should analyze them by statistical methods before interpreting them. The clarification of statistical tests is necessary to determine whether the differences observed between the studied groups are significant or not. Therefore, data analysis could not be a black box. It is essential to have an idea of the type of analysis envisaged when developing the study project. Analysis plans are an integral part of the research plan and must be clearly stated in the research design section of the proposal. The preparation of such plans allows the researcher to avoid several pitfalls. Otherwise, the researcher may discover at the end of the study that crucial information has not been gathered, or some of the collected data was not assembled in a suitable form for statistical analysis and cannot be included in the analysis.

The name of the software used for statistical analysis is usually mentioned. Most computer programs are equipped with Excel software. The most common software in medical science is probably Statistical Package for the Social Sciences (SPSS); however, there are other software packages such as MINITAB and GraphPad prism. These tools enable you to enter the raw data and perform a wide range of statistical and multivariate analyses on them.

The choice of statistical analysis depends on the type of data, their collection, and the hypotheses to be verified. Based on the research question, the hypothesis (relationship between variables or a difference between groups), and the type of data (quantitative or qualitative), you can decide about the required statistical tests and explain the rationale for your decision. This section should also describe the assessment of errors and the distribution of the variables.

Example

"For all aims requiring comparison of … we will compare samples by using … test via … software. We will report the mean differences, 95% confidence intervals, and p-values for descriptive purposes. We will focus hypothesis testing on … using a false discovery rate (FDR) threshold of 0.05. For aim, the primary endpoint is …, which we will compare via a generalized estimating equations (GEE) approach in linear regression. We assumed the minimum clinically meaningful effect size for in vivo studies would be … standard deviation units, which will require … animals per group with two-sided alpha = 0.05 and power of 80% under moderate correlation."

Briefly, you should ask yourself the following questions to select a correct statistical test:

- What was the research question? What are the main comparisons you need to make to answer the research question (summarize the hypotheses)?
- Do you want to evaluate the association between variables or the difference between groups (correlation or comparison)?
- Is your data qualitative (nominal) or quantitative (continuous)? Will the factors under study be expressed as a proportion or as measured values on a continuous scale?
- How many groups do you have? Will you be comparing two or more groups?
- Will you compare two or more measurements in a single group? Is one or two observations collected per sample for each variable or repeated observations (unpaired or paired)?
- Is the data distribution normal or non-normal?

4.6.5 Potential Pitfalls and Alternate Approaches

Potential limitations of the selected research's type and data collection methods should be considered in the proposal. Here, you provide the reasons why you think the limits related to your choice are acceptable. The suggestion of alternative plans shows that while making the case that the proposed approach is the most effective way to tackle the research question, the researcher considers that the study activities may not proceed as envisioned. Consider the following example for talking about potential pitfalls and alternative strategies (https://www.niaid.nih.gov/sites/default/files/1-R01-AI121500-01A1_Gordon_Application.pdf):

"The working hypothesis for this aim is that a biofilm's evasion of the immune system depends on both the mechanics of the biofilm and its production of virulence factors. This hypothesis is based on the literature and our own results, but it carries a measure of risk since very little is known about how biofilms evade the host immune system, and because these types of measurements are new and new protocols will have to be developed. If the agarose and polyacrylamide gels resist phagocytosis despite provoking an immune response, we will first test whether they are too stiff and/or have yield stress too high for neutrophils to overcome. To extend the range of mechanics assayed to softer, weaker materials, we will use alginate and other hydrogels with moduli that are tunable even lower [265–269]. Should we find that, even in this case, mechanics do not contribute to immune evasion, we would use AFM and shear flow to measure the strength of adhesive forces between neutrophils and biofilms with different polysaccharide production patterns".

4.7 Timetable

A milestone or Gantt chart is a table in which the researcher determines the beginning and end of executive activities and the length of time to perform the various stages of research. The researcher must realistically determine how the study will be applicable in a particular time process. For this purpose, it must provide a timeline appropriate to the type of research.

The timetable should be planned based on each phase of the research by considering:

– The pilot studies
– Preparation of materials and tools
– Synthesis of samples/sampling
– The data collection
– The statistical analyses
– The research reports

4.8 Budget and Personnel

The various planned expenses should be explained in detail by attaching a written budget description. The proposed budget may include indirect costs (facilities and administrative costs) and direct costs for supplies, purchased service, personnel, fringe benefit, and/or travel. Remember that equipment such as computers and administrative expenses such as secretarial support may not be credible for funding. You should also consider the budget justification by talking about the need for each item and the time in which the money will be spent [10].

Biographical sketches (bio-sketches) should provide a clear description of the qualifications and experience of the researchers, including their training, university degrees or credentials, research experience, and scientific publications in this specific area of research. You should also describe the personnel required to carry out the study and define each person's tasks. Other responsibilities and occupations of the researchers could also be found in this section to show the reviewers that researchers have enough time and sufficient experience to carry out the research [11].

4.9 Core Review Criteria

Reviewers will evaluate the proposal's weaknesses and strengths. To be ready for this step, it is essential to discuss your concept with program directors before writing the proposal. The next step after writing the proposal is to ask some colleagues to review it before submission. Finally, anticipating the reviewers' questions makes it possible to avoid errors and ensure that your research project impacts its reviewers positively. Ask yourself if the proposal tackles a significant problem, opens up new discoveries, or exerts a considerable influence in your field, and your specific aims are written clearly (easy to understand) (https://www.niaid.nih.gov/grants-contracts/draft-specific-aims). Concerning the proposed research, reviewers would consider the following questions:

- How original and logical are aims or questions, and to what extent are the hypotheses formulated clearly?
- Are the impacts made by the research project expected to be significant or original? What are the prospects of being able to acquire new knowledge of any importance?
- To what extent the proposed research is based on hypotheses or known issues? Are the proposed methodology and data analysis appropriate? Will the applicant and other potential researchers apply the new methods to be explored or developed? Did the candidates anticipate difficulties, and, if so, what research solutions are proposed?
- To what extent has the relevant scientific literature been reviewed and assessed? Does the researcher formulate the theoretical and conceptual framework for the proposed study?
- Do the applicants have the necessary training or experience to carry out the proposed research? How essential and original is the applicants' previous work? To what extent can applicants complete the proposed study considering various factors, including the amount of time they plan to spend on the research?

Almost all NIH-funded research is subject to peer review. The reviewers would determine whether the projects presented go in the direction of the NIH mission, which is the advancement of knowledge in medicine and the biomedical field. Since much of this information will be provided in specific aims on first reading, developing specific aims is critical. The first part that the reviewer will read is specific aims, and based on that, he/she will judge the other parts. A fishing expedition, ambitious, correlative, descriptive aim, not-feasible hypothesis, lack of literature review, not innovative work, or low impact are the common critiques reported about the specific aims page [12]. Three Ps, which are personnel, proposal, and place, will be considered during proposal review [13]. More specifically, five criteria will be investigated: significance, investigators, innovation, approach, and environment.

1. In the *significance* section, the reviewer will consider if you addressed a critical problem that advances science, explained its compelling rationale, and mentioned its relevance or future impact (in terms of improving scientific knowledge, technical capability, or clinical practice). When you have an important question, your proposal should clearly address the existing gaps or conflicts in the knowledge of the problem, how you can reduce the gaps by your proposed research, and the impact of your proposed solution. A pilot study can show the effectiveness of your theory or intervention. The specific aims should be focused on, the expected outcome of each aim should be described, and the future direction should be clear. Briefly, the most critical questions that reviewers will ask in this aspect are: Does the topic of study relate to a significant problem? If the objective pursued is achieved, how will scientific knowledge be more advanced? What will be the impact of this work on the concepts or methods that dominate the field?

2. The research *approach* must be justified, appropriate for testing the hypothesis, and suitable for reaching the research objectives. Additionally, the feasibility of each specific aim should be shown. Therefore, you should take the time to develop preliminary data for each aim. Incorporating enough preliminary data would establish the feasibility and support the hypothesis. Using your own preliminary data, also demonstrate your expertise. Furthermore, this section should include enough details about control groups, what data will be collected, how they will be collected, how they will be analyzed, the anticipated results, potential pitfalls, and alternate plans. Briefly, the reviewers will consider the following questions in this regard: Are the presented conceptual framework, the design, the methods, and the analysis adequately articulated, well-integrated, and in accordance with the objectives of the project? Does the researcher recognize the existence of sources of problems and consider an alternative strategy?

3. The reviewer seeks to find *innovation*, novelty, and originality in the project's idea, approaches, instrumentation, or interventions. New conception may concern a theoretical framework, a methodology, a process, a technique, a product, etc. The reviewers consider seeing if the project calls for concepts, approaches, or methods presenting innovative character. Are the pursued goals original and innovative? Does the project challenge the established paradigms? Does it develop new methodologies or new technologies?

4. Other essential elements for reviewers are the *principal investigator's* qualifications, the relevance of the number of team members, and the potential interdisciplinary synergy. The principal investigator and other team members' experience, research skills, productivity, and academic background concerning the project

will be evaluated. Therefore, you should demonstrate your training and experience and one of the other members of your team in your application to convey your capacity to conduct the proposed research. In the preliminary study sections and bio-sketches, you can highlight the relevant work and each team member's role in the proposed study. The factors that reviewers will consider are the applicants' knowledge, expertise, and experience to carry out the project; past or potential contributions to the proposed research area and their impact; and the complementarity of the team members' expertise and their synergy.

5. Finally, institutional support, resources, research centers, labs, and equipment, including the research environment, have a significant role in the project's success. Is the scientific environment in which the work will be carried out favorable? Will the proposed experiments take advantage of this environment? What is the institutional commitment to the applicants, and what impact does the group have on the institutions concerned in terms of the institution's research priorities, financial support, time devoted to project, and laboratory equipment and space? Will the institution support the project?

Based on the above criteria for NIH proposals, three reviewers will give your application an overall impact score of 1 (best) to 9 (worst). Your application will not be discussed if it is scored above the threshold of impact scores (poor). In case its score is below the threshold, it will be discussed and scored in the panel.

The National Science Foundation (NSF) is another agency founded to promote the advancement of science and place more emphasis on multidisciplinary activities. The NSF also has based its research support decisions on the peer review system. The evaluators judge the relative weight of two critical criteria according to the context and rank the proposals according to five categories, from "excellent" to "poor." These two criteria that frame the work of the reviewers must answer two questions [14]:

1. What is the intellectual merit of the proposed project? The reviewer will consider the importance of the project concerning the advancement of knowledge in his/her field or across different areas, the qualifications of the researcher or team of researchers by commenting on their previous work, the potential of the project to explore fruitful and creative concepts, the design and organization of the project, and access to resources.

2. What are the broader impacts of the proposed activity? The reviewer will consider the extent to which the project allows the advancement of knowledge while promoting teaching/training, whether the proposed approach can increase the participation of certain underrepresented groups in science and engineering (women, minorities, etc.), whether the project improves infrastructure for research and education, whether the results are likely to be widely disseminated, and what the benefits of the project for society will be.

Remember that the reviewers' comments are not about you but about the material you included in your proposal. In your revised application, you must acknowledge the help of reviewers' comments and address all the criticisms. You can make

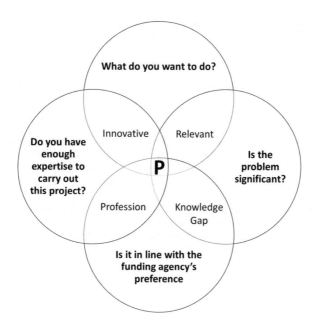

Fig. 4.6 The elements to be considered for a successful proposal (P)

your proposal interesting and informative by reviewing template proposals, brainstorming with experts or colleagues, storytelling with your preliminary data, and presenting it in an acceptable format while ensuring it is easy to read. A well-prepared proposal is the judicious integration of systematic development of an exciting and relevant problem that is critically needed to be resolved, along with the researcher's expertise and the funding agency's preferences. By preparing the proposal at the intersection of the four circles (Fig. 4.6), the reviewer will be convinced to be your advocate.

References

1. A. Gholipour, E.Y. Lee, S.K. Warfield, The anatomy and art of writing a successful grant application: A practical step-by-step approach. Pediatr. Radiol. **44**(12), 1512–1517 (2014)
2. L.S. Marshall, Research commentary: Grant writing: Part I first things first …. J. Radiol. Nurs. **31**(4), 154–155 (2012)
3. E.K. Proctor, B.J. Powell, A.A. Baumann, A.M. Hamilton, R.L. Santens, Writing implementation research grant proposals: Ten key ingredients. Implement. Sci. **7**(1), 96 (2012)
4. K.C. Chung, M.J. Shauver, Fundamental principles of writing a successful grant proposal. J. Hand Surg. Am. **33**(4), 566–572 (2008)
5. A.A. Monte, A.M. Libby, Introduction to the specific aims page of a grant proposal. Kline JA, editor. Acad. Emerg. Med. **25**(9), 1042–1047 (2018)
6. P. Kan, M.R. Levitt, W.J. Mack, R.M. Starke, K.N. Sheth, F.C. Albuquerque, et al., National Institutes of Health grant opportunities for the neurointerventionalist: Preparation and choosing the right mechanism. J. Neurointerv. Surg. **13**(3), 287–289 (2021)

7. A.M. Goldstein, S. Balaji, A.A. Ghaferi, A. Gosain, M. Maggard-Gibbons, B. Zuckerbraun, et al., An algorithmic approach to an impactful specific aims page. Surgery **169**(4), 816–820 (2021)
8. S. Engberg, D.Z. Bliss, Writing a grant proposal—Part 1. J. Wound Ostomy Cont. Nurs. **32**(3), 157–162 (2005)
9. D.Z. Bliss, K. Savik, Writing a grant proposal—Part 2. J. Wound Ostomy Cont. Nurs. **32**(4), 226–229 (2005)
10. D.Z. Bliss, Writing a grant proposal—Part 6. J. Wound Ostomy Cont. Nurs. **32**(6), 365–367 (2005)
11. J.C. Liu, M.A. Pynnonen, M. St John, E.L. Rosenthal, M.E. Couch, C.E. Schmalbach, Grant-writing pearls and pitfalls. Otolaryngol. Neck. Surg. **154**(2), 226–232 (2016)
12. R.J. Santen, E.J. Barrett, H.M. Siragy, L.S. Farhi, L. Fishbein, R.M. Carey, The jewel in the crown: Specific aims section of investigator-initiated grant proposals. J. Endocr. Soc. **1**(9), 1194–1202 (2017)
13. O.J. Arthurs, Think it through first: Questions to consider in writing a successful grant application. Pediatr. Radiol. **44**(12), 1507–1511 (2014)
14. M. Monavarian, Basics of scientific and technical writing. MRS Bull. **46**(3), 284–286 (2021)

Additional Resources

https://grants.nih.gov
https://grants.nih.gov/grants/oer.htm
https://www.ninr.nih.gov
https://www.niaid.nih.gov
http://www.grantcentral.com
http://www.saem.org/research
http://www.cfda.gov
http://www.ahrq.gov
http://www.nsf/gov

Chapter 5
Scientific/Clinical Research Report

5.1 When Should You Start?

Research results should usually be presented to the institute or university as a dissertation or an article. A dissertation (thesis) should be written according to the policy of each university. In general, the dissertation is a detailed article that deals with each part of the article comprehensively so that other students can get practical information by referring to that dissertation. Hence, dissertation chapters are almost like articles and include Introduction, Materials and Methods, Results, Discussion and Conclusion. For more details about dissertation preparation, you can refer to the relevant university's *format* requirements. For reframing your dissertation to a manuscript, you should not "copy and paste" from your thesis but instead rewrite phrases after selecting. Table 5.1 summarizes the different sections of the research paper.

If we show the different stages of research in a step-by-step manner, writing is the last step. Although writing a good scientific paper takes time, if you start writing at the same time as you begin to think about a research problem, you will have less trouble in preparing your manuscript. Writing is a continuous part of the research process that should commence immediately after determining the research topic and continue during the research until the end of the study. The researcher starts taking notes from the observation, thinking, and reading steps then prepares the research proposal. While performing the research, he/she regularly writes down the assays and observations to understand his/her mistakes and be able to repeat the work more carefully later. Finally, he/she must document the results in the form of a report, dissertation, or article. Remember that systematic and regular writing from the beginning of the research will assist you in writing the final report. Taking notes during the literature review will support the "Introduction" and "Discussion" sections. Correspondingly, preparing a proposal could be very helpful in manuscript writing. The proposal's statement of the problem, significance, and specific aims will be summarized in the "Introduction" of the future article. The proposal's approach will be included in the "Materials and Method" portion of the manuscript.

© Springer Nature Switzerland AG 2022
F. Tabatabaei, L. Tayebi, *Research Methods in Dentistry*,
https://doi.org/10.1007/978-3-030-98028-3_5

Table 5.1 Different sections of a research paper

Section	The question you should answer in this section	Objective
Introduction	Why did you start this research?	A summary of the research background, literature review, your objective
Materials and Methods	What did you do in this research?	Provide sufficient details on how to conduct the research to allow for repetition
Results	What data did you get?	Use data analysis to respond to the research question
Tables and Figures	What do the results show?	Express results more clearly
Discussion/ Conclusion	What do your results mean?	Interpretation of research findings in comparison with other published articles in the same field and identification of possible clinical application of the obtained results
References	Who else has done significant research in your field of study?	References to the latest and most reliable published articles

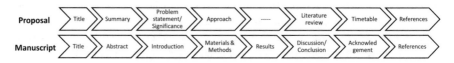

Fig. 5.1 Comparison of the different sections of proposal and manuscript (Literature review in proposals for NIH grants is usually within the Significance section or justification of each aim in the Approach section)

Likewise, writing down the details of your methods and tests while performing the research would help you complete the "Materials and Methods" section. Finally, the proposal's literature review will be considered in the "Discussion." The only part not appearing in the proposal is the "Results" since the study has not yet been carried out (Fig. 5.1). You can quickly prepare the manuscript's initial draft by putting everything, including your notes from the literature review, your daily lab notebook containing the details of all assays and observations, and the research findings, in a Word file. There is no need to correct the sentences at this stage. If you do not have information about a part, leave that page blank and go to the next page. You can come back later and modify the first draft.

In writing a scientific paper, the content characteristics, as well as the methodological structure, should be considered seriously. In other words, although using the best materials for delivering a product to the market is essential, you should also know how to attract customers to this new product. The content of the article should be based on general principles so that they can be quickly reviewed. If an article does not have these principles, it cannot be considered for a scientific evaluation. A logical and straightforward arrangement (i.e., introduction, materials and methods, results, discussion, and conclusion) can provide point-by-point answers to the problem. In this regard, while scientific articles are accessible at different levels, their evaluation and review are often based on a series of specific standards, most of which are common. In addition to the fact that the article should have the main structures, how these parts relate is also essential. Therefore, it is better to follow the principles of paragraph writing in each of the article's sections.

The main question here is how to write a scientific paragraph for an article. A paragraph is a collection of several sentences that describe the core concept. Typically, the first sentence of the paragraph refers to what is known and conveys important information about the specific topic. The following sentences should provide the supporting information and link them with transitions. The last sentence of the paragraph summarizes the main point or makes an important point about what is unknown. The paragraph is related to the next paragraph by the final sentence. In writing all these sentences, the author must articulate and explain a single concept in a logical order. Therefore, a well-organized paragraph has three main parts: (1) the subject of the paragraph or specific topic that includes the opening/topic sentence and is the center of attention (introduction), (2) supporting sentences that provide more information and expand the specific topic (body), and (3) final sentence which indicates that the paragraph has ended and reminds you of an important point that you must remember (conclusion). If we look at the paragraph's structure more closely, it is like a short article that contains 100–400 words. The topic sentence can be considered an introduction, supporting sentences are the body, and the final sentence is like the article's conclusion. Therefore, a paragraph is a "multi-line article," and the final manuscript will be excellent if all these small articles are written correctly.

The scientific paper should not only be structured but also be precise and straightforward. The researcher should talk about the results that have the required validity and specific research characteristics. Scientific precautions prompt the researcher to avoid exaggerating and provide brief and convenient scientific material based on the research purpose. Hence, writing is not just putting words together; it is a kind of art to bring thought to a beautiful, complex, creative process in which the author skillfully combines different words to create a meaningful connection with his/her mind and communicate that message to others. The transfer of scientific concepts is the primary role of scientific articles. Consequently, it should be written so that the reader can convey these concepts with the least amount of time and energy. Achieving this goal is possible by:

- Revising long sentences (>30 words)
- Using short and concrete words and action verbs instead of nouns (improvement, identification, function)
- Reducing the use of pronouns and acronyms
- Avoiding ambiguous words (such as "important," which is only qualitative) or negatives (not negligible = important)
- Deleting imprecise words (really, absolutely, incredibly)
- Building the sentence around an action verb

But how long does it take to write the report of a research project? Usually, after performing the research, you should consider 2–4 months for evaluation and analysis of data. Each step of writing and submitting the manuscript, evaluation by reviewers, response to reviewers, proofreading, and publishing the article would also take 2–6 months, depending on the collaboration of team members and journal speed. Therefore, the publication process (from submission to acceptance of the manuscript) could take 6–12 months or more.

We have used the terms "manuscript" and "article" interchangeably, but usually, the scientific paper that has not been published is called *manuscript*, while the *article* has been published. We will point out how to write a manuscript and prepare it for publication in the following sections. Remember that "the hardest part is getting started." So, just get started. If you know the rules and follow them, the subsequent steps will be easy.

5.2 Title

The title of a scientific article is the first contact with reviewers, readers, search engines, databases, and Internet users. An attractive, informative title must be written with clarity and completeness, allowing its comprehension without recourse to the full text of the manuscript. Begin the title of your manuscript with the most important words, which depend on your objective, and focus it on the actual results or new idea carried out by your research—example: "Improved dental adhesive formulations based on reactive nanogel additives" [1].

Try to avoid abbreviations, parentheses, and unnecessary words such as "Study of" and "Analysis of the effect of" in the title. You can skim the journal in which you wish to publish your manuscript to be more familiar with the style of the title of articles already published in that journal. The *instructions for authors* of scientific journals may set some limitations for the length of the title. We have already described the characteristics of a good title (Chap. 4). It may contain the PICO elements; in that case it is indicative or informative. For example, "Musculoskeletal symptoms (Outcome) among dentists (Population) in relation to work posture (Factors studied)" is an indicative title [2]. In contrast, "Work-related (Factors studied) musculoskeletal disorders (Outcome) in dental students (Population): a cross-sectional (Study design), pilot study from a UK university teaching hospital (Location)" is an informative title [3]. The title of your manuscript can be the same as the title of your proposal. It is recommended to start or end the title of the review article with a word that shows it is a review (a review, an overview, etc.)—example: "Antimicrobial Therapeutics in Regenerative Endodontics: A Scoping Review" [4].

Usually, journals require you to submit the title page, which includes the authors' names and affiliations, separately from the rest of the manuscript.

Some journals (e.g., *Tissue and Cell*) entail the *highlights* or a *significance statement* that completes the information given by the title and summarizes the article's message. The highlights or significance statement may include the hypothesis, the variables studied, the methodology, the emphasized result, the implications, or the limits. You should summarize the main points of your study, which shows the originality and interest of your research and should be understood by readers without reading the abstract or the text of the article.

Example of highlights provided in a study with the title of "In situ surface modification on dental composite resin using 2-methacryloyloxyethyl phosphorylcholine polymer for controlling plaque formation" [5]:

- "Antibiofouling nature of dental composite resin is important to prevent infection.
- Dental composite resin surface could be modified with photoreactive MPC polymer.
- A procedure like dental treatment in situ could be applied for the modification.
- Protein adsorption and bacterial adhesion were dramatically reduced by the modification".

Some other journals (like *Operative Dentistry* or *Tissue Engineering*) may require *clinical relevance* or *impact statement* (120 words or less), which is about the novelty or importance of your work with respect to the field of study or the society.

5.3 Abstract

The abstract is a precise reflection of the content of the manuscript. It is freely accessible and briefly describes the research's scope and objective, how it was done, its findings, and conclusions. The purpose of the abstract is to provide a first glimpse of the research structure in a short and straightforward form. It will guide other researchers, save them time collecting and selecting information, and assist them in research and decision-making. Most of the readers interested in the article's title would read the abstract to decide whether to read the full text. Therefore, the abstract should be concise while being sufficient for the reader to understand the study's primary objective, the approach taken to carry it out, and the results.

Abstract writing means recognizing the essence of a text and presenting it concisely with sufficient accuracy and clarity. The main contents of a project should be identified before writing the abstract. Usually, the first sentence of the abstract explains the need to address an issue and the research objective. Then, methods of performing the project, findings, and conclusion are presented. In other words, the abstract should include the critical question of the study, the work done to find the answer to the research question, the data obtained, and the answer to the research question. There is no need to describe the main problem, review the literature, give a citation, and discuss the results in the abstract.

Writing a correct, accurate, and valuable abstract depends on the following principles:

1. The abstract should reflect the article and provide a clear and concise picture of the project. Extract the important and main content from the original text and present it in short sentences.
2. The readers should understand the abstract without referring to the full text. Avoid using jargon words and vague phrases.
3. The length of the abstract can be between 150 and 350 words. To find out the number of abstract words, you can select the entire abstract. At the bottom of the word page (left), the number of words chosen and the total number of words in your file will be displayed.
4. Write the abstract with the same structure and order as you wrote the article; start with a summary of the introduction, then give a brief overview of the method, results, and conclusion.
5. Each *structural abstract* includes four main headings: Objective/Background, Methods, Results, and Conclusions. Some journals (like *Acta Biomaterialia*) require a *non-structured abstract*, in which you don't need to mention the headings in the abstract. Therefore, in non-structural abstracts, the abstract should be written in one paragraph while maintaining harmonization between sentences.
6. In the structural abstract for an *original article*, answer these questions in each section:
   Background and objective: "What is the context in which your research takes place?"
   Materials and methods: "How did you do it?"
   Results: "What are your findings?"
   Conclusion: "What do the results mean scientifically?"
7. In the structural abstract for a *review article*, answer the following questions:
   Background and objective: "What is the PICO question that you are exploring?" and "Why is it significant?"
   Materials and methods: "What are the sources used, search strategy, method of quality assessment, and data extraction?"
   Results: "What major points emerge from your review?"
   Conclusion: "What do you highlight: contradictions, interpretations, new questions?"
8. Do not neglect the research question, do not state the question vaguely. For the objective section of your abstract, you should mirror the last paragraph of the Introduction.
9. In the results, it is better to state the mean ± standard deviation and the number of samples along with the *p*-value.
10. The conclusion should be justifiable and in response to the main question. It must be supported by the data you are presenting. Do not state the application instead of answering the question.

11. Avoid referring to parts of the article, brand names, table/figure, or references, and do not use footnotes in the abstract.
12. Sentences should be consistent, coherent, and logical and should not be separated or fragmented. Observe grammatical and editorial points and the coordination and uniformity of verb tenses (past or present tense). Use the present verbs for research question/answer and use past tenses for methods and results. Avoid using a question mark (?) and the exclamation mark (!) in the abstract.
13. Favor the use of "we," "our," "explicit agents," and active verbs when possible.
14. Write the other parts of the manuscript and go to the abstract last. It is placed at the head of the manuscript but is often not written until the whole manuscript has been finalized.
15. Look at some abstracts in specialized journals and use them as a template. Most journals indicate how the abstract should be written and the number of words to be counted in the *authors' instructions*.

Some journals (e.g., *Materials Science and Engineering C*) request a graphical abstract that illustrates the critical result or the process of conducting research (Fig. 5.2). As a graphical abstract usually has no legend, it should be understandable on its own. Instead of reusing one of the figures of your manuscript, you can create a figure based on the title of your paper.

Usually, three to six keywords that describe the content of the article are expressed after the abstract. These keywords will be employed by search engines (especially Google Scholar) and databases (PubMed, Scopus, Web of Science) to reference the article. For keywords:

• Choose words that you hope readers will use to access your article (concept, method, element, implication, abbreviation widely used in a discipline).

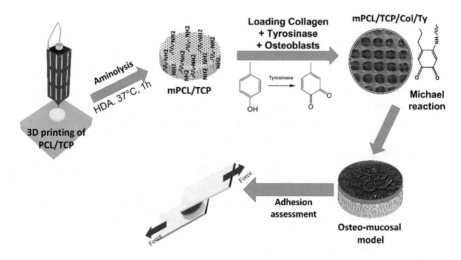

Fig. 5.2 An example of graphical abstract [6]

- Use specific words that highlight the central message of your article and make it quickly found on the web.
- Select the essential words not used in the title but used in the abstract and their widely used synonyms.
- The repetition of keywords (in the title + abstract + keywords) helps the first occurrences of your article in the search engine or the database. But avoid overdoing repetitions.
- Words related to a broader field would allow the discovery of your article by interdisciplinary readers.
- To evaluate the usefulness of your keywords, you can try them in search engines and databases.
- It is often recommended that the provided keywords be found in the MeSH list.

You should carefully think about how to prepare the abstract. In many cases, the abstract of a submitted manuscript is the only part that editors will read. A well-written abstract could play an essential role in editors' decision to reject the manuscript or send it for peer review. You can prepare a checklist for your abstract section to see if the following details are specified:

- Background and purpose, methods, results, and conclusions are provided.
- Key specifications are stated.
- The research novelty is indicated.
- In the method section, the type of study is stated.
- The abbreviations used in summary are spelled out.
- The conclusion is justifiable and based on the research question.

5.4 Introduction

Each scientific article begins with an introduction, which is a summary of the first and second chapters of the dissertation, outlines the research question, and reviews its background. An introduction is an answer to why you chose this research topic. Here, you must talk about the research problem, the need to perform the research, the studies directly related to your research subject, and your objectives. The introduction should provide a fundamental logic to the study and show the reader why this research is a logical continuation of previous studies. For preparing the manuscript's "introduction," you can use "the problem statement," "literature review," and "specific aims" sections of your proposal.

Answer a few basic questions before writing the introduction: What should be mentioned in this research introduction? What is the correct way to express it? Who are the readers of this manuscript?

The introduction—like a window—takes the reader from the outside world to the world inside the article. The readers in this section are looking to find answers to three main questions: What do we know about the research topic? What do we not know? And why was this study done? The same for a review article, your introduction answers three key questions: What subject have you chosen to explore in-depth by analyzing the published results? Why is this subject interesting in the current context of research questions in your field? What significant points emerge in terms of research questions and implications?

A logical and scientific process is necessary to answer these questions. The course of presentation in the introduction is from the general (known) to the specific (objective). The authors begin the introduction with a general explanation of the problem in an interest-catching way. They gradually move on to more explicit sections like a review of the problem's background and the gaps in current knowledge that this study intends to fill. Finally, they clarify the need to complete this gap which is the study's objective or aim. Therefore, in the introduction:

1. You should first have an overview of the research topic and give a broad background of the subject of the study.
2. It is then necessary to concentrate on the topic and provide more information about the research question. You briefly review the literature published in this field and identify the already known things about the subject.
3. Then it is time to highlight the remaining questions and the problem that echoes a lack of knowledge in the scientific literature.
4. Finally, you should state the purpose of your research which must be formulated clearly and follow logically from the problem.

Write for your audience, not for yourself. You should know the interest of the readers of your manuscript. Sometimes authors begin the introduction with an idea and hope that the reader will understand its points with personal wisdom. Some other times, authors include trivial information that cannot add anything to the readers' knowledge.

There are more details in the introduction that need to be followed. For example:

- Always pay attention to the compatibility of the title of your manuscript and the problem being studied. Do not talk about content that is not relevant to your research.
- Clarify in the first paragraph what the problem is but avoid going too far into the topic's history.
- Methodology, results, and discussion should not be declared in the introduction.
- Avoid meaningless acronyms or unfamiliar abbreviations as much as possible. Be sure to explain the full names the first time you use abbreviations/acronyms in the manuscript.
- Try to define the specific word, illnesses, or medications. Nothing can stop the readers from reading your work faster than references to terms they do not know anything about.
- The introduction should not be written as a review article. It is necessary to mention the previous studies without interpreting them (we will do it in the discussion by comparing our results with those of the other researchers).
- Use a systematic way to introduce work done by other researchers and highlight the discrepancy in the results or the gap in the knowledge: "Using universal dental adhesive, Pfifer et al. found... Silva et al., however, failed to replicate those results... In summary, the above work shows...."

- If you want to quote a phrase exactly from others, put it in the quotation marks.
- Avoid repetition of keywords because they damage the beauty of the introduction. You can use frequent synonyms for imperative words to make it easier for your manuscript to be found through search engines.
- A lengthy introduction shows that we do not know how to formulate the problem. The introduction should not exceed 10% or one-sixth of the entire article (10–20 references, 3–5 paragraphs). However, the review article takes the form of a longer research article. *Dental Materials* impose a maximum length of six journal pages for original research reports, while this limit is ten pages for review manuscripts.
- The objective statement is the last sentence of the introduction that should describe why the reader should care about your experience: "The objective of this research is to determine…."
- Try to highlight the importance of your research while avoiding direct statements like "This is the first report of…." Never claim that you are the first person to write in this field.
- Use simple present tense when referring to a subject that has been accepted and apply past tense or present perfect tense when referring to previous research and studies. Present-perfect tense usually comes with the words "recently" or "currently." Example: "GICs *have* unfavorable characteristics that limit their clinical application. Lucas et al. *reported* that when hydroxyapatite was added to a GIC, the material demonstrated a crystalline structure similar to that of human dental structures, enhancing the mechanical properties of the cement. *Recently*, a new type of glass ionomer, EQUIA Forte Fil (GC Corp), *has been introduced*" [7].

You can use a checklist for your Introduction:

- What problem is addressed?
- Is it clear why you did this study and why the subject of your manuscript is significant?
- Did you mention the most important articles related to your field of research?
- What does your study add to existing knowledge?
- Is the critique you have given to previous studies justifiable?
- Is the main research question formulated precisely (PICOT)?
- What parts can you delete?

5.5 Materials and Methods

This section is the most straightforward section of the manuscript to write. Depending on the journal, it could be positioned just after the introduction or at the end of the article. The purpose of this section is to describe how a study was conducted. It should be detailed enough so that readers can reproduce the experiment. This section also lets readers judge the quality of your work and criticize the validity of the results. Consequently, it would be better to know and respect the standards of reporting materials and methods based on your study type. For example, you can adhere to Strengthening the Reporting of Observational Studies in Epidemiology (STROBE) for an observational study [8], ARRIVE for animal research [9], CONSORT for a randomized controlled trial [10], Standards for Reporting of Diagnostic Accuracy (STARD) for a diagnostic study [11], or PRISMA for a systematic review [12].

The type of study (cross-sectional, cohort, clinical trial, or experimental) should be specified. Sometimes it may be helpful to present the study design as a figure. Structuring the "Materials and Methods" section based on the experiments performed is also very helpful. Methods used for sample size calculation, randomization, random assignment, and blinding should be explained for each test.

The method of doing the work should be mentioned in detail so that other researchers can evaluate or reproduce the tests or experiments of your study. If the employed technique is very known, a brief description is sufficient, giving the references which allow the detailed description to be found. If another researcher's method has been used in a procedure, it should be referred to that researcher. If you are modifying a previous approach, refer to it, and describe the modifications. And if the method of work is entirely innovative, it is necessary to mention all the details. If previous studies used a technique like yours, cite them to give more credibility to your research methodology.

All the materials can be collected in a table with details about the ingredients, the manufacturer, the country, etc. In the case of devices, it is necessary to mention the generic name, the manufacturer, the name of the city, the country, and sometimes the lot number in the parentheses. Also, the accuracy of devices such as micrometers, scales, etc., should be mentioned. Example: "weights were determined to be ± 0.0001 g using a digital balance."

Human or animal studies require the permission of the ethics committee, informed consent from patients, and the age and sex of the specimens. Patients' identity information should not be revealed, and it is mandatory to cover the eyes and parts of the patient that indicate their identity (Helsinki Treaty). Sampling, random assignment (experimental and control groups), details of interventions performed for each group, definition of efficiency and how to measure it, methods to increase the quality of measurements, and the process of stopping the effect of interventions should also be explained in this section. The materials and methods section answer the question of on which sample or on whom the study was performed. What is the sample size, and what is the justification for this sample size? How is the study done? Where and when was it done? By what means was it done? And what were the inclusion and exclusion criteria?

At the end of this section, you should describe the statistical methods used to analyze quantitative data (does not apply to qualitative data) and specify the variables concerned for each. This section enables a reader with access to the original data to verify your results. The reproducibility of the results is a guarantee of scientific honesty. The experimental setup (designing the proposal) and the statistical analysis are closely linked. It is the experimental setup that guides the analysis. The decisions taken during the design phase of your research must be considered in the choice of statistical methods. Data can be reported as mean ± standard deviation (SD), mean ± standard error (SE), or mean ± confidence interval (CI). In the case of normal distribution, SD has an excellent property to show data variation. If more than one method has been used for data analysis, they should be disclosed so that readers can make their own judgments. The significance of the p-value, which is usually less than 0.05, should be mentioned. It is also essential to specify the software and the version used to perform this analysis. Finally, if a biostatistician did the statistical analysis, it can be noted in the acknowledgment of the manuscript.

You can use present verbs to explain how data is presented (e.g., data *are* summarized as means ± SD). Use past tense (passive voice) to report how work is done. For example: "scanning electron microscopy (SEM) and energy dispersive spectroscopy (EDS) *were performed* to assess the microstructural changes caused by the solutions on both dentin and sealer ($n = 3$)" [13].

You can prepare a checklist for your "Materials and Methods" section to see if the following details are specified:

- Research design
- Sample size, detailed specifications of samples, and how to assign groups
- Inclusion and exclusion criteria
- Ethical requirements (the permission of the ethics committee, otherwise notify that the research complied with the Declaration of Helsinki)
- Materials, measuring tools, methods, interventions
- Statistical analysis (software, tests, p-value)
  Can the reader repeat your study based on the given details?

5.6 Results

This section should answer the question: What have you achieved from this research? In this section, based on the data analysis, you will express the findings of your study. The purpose of research should not be confused with the result. The result is about expressing the collected data, while the research purpose is to consider the null and the alternative hypothesis. To achieve the goal of the Results section:

1. Identify the hypothesis to which each analysis refers. Suppose you have used various experiments and reached different data. In that case, it is better to create separate subsections in the Results section and report the results of each test in a particular area.
2. Report all the results, including those that go against your hypothesis.

3. Report the exact p-value (e.g., $p = 0.001$ instead of $p < 0.05$) along with confidence interval (CI) to help readers to make their own judgments about how significant your results are.
4. Report the impact of a particular treatment and the average treatment performance level individually while comparing it to another treatment. Reporting the effect size and its precision (confidence interval) could be beneficial.
5. Respect SI writing conventions (e.g., spaces between numbers and units, parameters in italics, functions in roman).
6. Avoid interpretation in the Results section. In this section, you should only present your findings neutrally and objectively. However, if the journal requires merging the Results and Discussion, the interpretation should be written immediately after the description of each result.
7. Avoid referring to other articles in the description of your results unless you interpret the result when the discussion is merged with this section.
8. Summarize changes in time, morphology, differences, and similarities using the verbs "increase," "decrease," "greater," and "smaller."
9. All data interpreted in the Discussion should be mentioned in the Results section.

Using tables and figures in a report or research paper is a good idea if they are relevant to the subject and legally permissible. However, values that are already presented in figures and tables should not be repeated in the text. Images are in the form of graphs (diagrams), shapes, or flowcharts (processes). Usually, the table introduces a large amount of data, while the figure shows meaningful findings. In this section, you should also avoid duplication of data. A table enriches your subject with specific data, the development of which in the body of your manuscript would weigh down and hinder reading. If the results are clearly shown in the table, there is no need to repeat the descriptions in the text or display them in the figures. If statistical data are presented in the tables, it is necessary to mention the mean ± standard deviation. Tables should be as simple as possible. Two or three small tables are better than one large table with too many details or variables. In general, three variables are a maximum for easy reading. Each row and column should have clear and concise titles. Units of measure for the data must be specified.

Figures are diagrams, curves, histograms, or photographs and have a substantial and immediate visual impact. Correct preparation of the figures allows the reader to understand the overall meaning of the data quickly. Some presentation methods require more complete data than others. You should define which precise message you want to communicate then choose the method. Based on your data and the intended purpose, you can select between curves, bar charts, scatter plots, etc. In the *graphs* illustrating quantitative data using a coordinate system, the vertical axis (y) shows the outcome/dependent variable, and the horizontal axis (x) determines the exposure/independent variable. Each axis must be named with the unit of measure in the parenthesis. The curves of the graph must be thicker than the coordinate lines. The scale intervals should be visible, as should the unit of division of the scale. Major ticks should be numbered, and the symbol should be explained. A scale break can be used with this type of graph, but in this case, care must be taken to avoid any misinterpretation. A *scatter plot* is a particular type of graph that shows the relationships or associations between two variables by using single points. In this chart type, pairs of related data series are represented by dots on the same chart. The resulting plot highlights a possible relationship. If it tends to follow a straight line, then the relationship is linear. If the plotted points are scattered, it can be inferred that there is probably no relationship. *Diagrams* are symbolic modes of presentation of statistical data using only a single coordinate. They are ideal for presenting comparative data. *Bar charts* showing the data in separate columns can be horizontal or vertical. It is best to arrange these bars in ascending or descending order for ease of reading. Never use a scale break in a bar chart; this will result in misinterpretation. Columns can be shaded, hatched, or colored to highlight the differences between the bars. In the case of comparisons, the space between bars in the same group is optional; on the other hand, it is obligatory between the different groups of bars. Error bars showing SD (for descriptive analysis), SE, or 95% CI (for statistical inference) should be clearly labeled. Headings should fully identify the data presented. All other indications should be clear and easy to understand. By checking the properties of image files, you can find the resolution information. Remember that a resolution of at least 300 dpi is required for correct printing. However, higher resolutions (600–1200 dpi) are usually requested by journals (instructions to authors). Color photocopies are expensive. The colors can be replaced by hatched or dotted surfaces or dashes (of different lengths on the various curves). Finally, you should provide the image files in the correct format (TIFF or JPEG formats are preferred).

All tables and figures should be self-explanatory and independently comprehensible (without reading the manuscript's body). A brief and concise explanation should be given at the top of the tables or below the images (legend). Abbreviations or symbols should be explained in detail in the image's legend or the table's footnote. Where you think it is appropriate, refer to the tables and figures separately in the manuscript's body (in the order of appearance). If you do not refer, there will be no reason for the reader to look at them. For this purpose, each legend must be numbered (example: Figure 3. Scanning electron microscope image of stem cells seeded on the scaffolds in the third day (scale bur …)). The numbering of tables and figures is independent.

Remember that you are not allowed to use copyrighted images. Therefore, if you want to use them in your article, you must have obtained written permission from the journal and the author of the published article. The name or names of the authors must appear after the legend in parentheses. Please note modification of a copyrighted image, even slight, is subject to the prior authorization of its author. The journal from which the figure was taken must appear in the bibliographical references. The objective is to respect the intellectual property of authors.

The results are written in the past tense, except when you are referring to figures or tables (example: "All groups *presented* high counts of periodontal pathogens at baseline. Figure 2 *shows* changes in the proportions of microbial complexes" [14]).

There are more details in the Results section that need to be followed. For example:

- Start with positive findings.
- Compare groups.

- Use a mixture of theoretical explanations, tables, and figures.
- Enter the number of subjects/samples, and specify the unit of measurement.
- Explain the signs, stars, and abbreviations in the figure's legend or at the bottom of the table.
- Present your results in the same order as the presentation of the methods.

The checklist for your Results section should consider the following details:

- The original data are expressed.
- The presented results are clear and understandable.
- The standard deviations are mentioned.
- Enough tables and figures have been used; all tables and figures have an informative title or legend and are numbered; tables and figures are referred to in the manuscript.
- All the abbreviations used in the figures and tables are explained in the legends.
- Health Insurance Portability and Accountability Act of 1996 (HIPAA) issues/patient identifiers are considered.

5.7 Discussion and Conclusion

One of the most important things in writing a research report is establishing the interpretation and arguments in the discussion. As we explained in Chap. 1 (Sect. 1.2), the researcher starts research by careful observation and creative thinking. In the final steps of the research project, the Discussion requires an effort of reflection again. After presenting data in the "Results section" using texts, tables, and figures, the researcher should now explain the meaning of the results in the study context by linking them to previous related studies.

The evaluation of the results in achieving the goal is based on the hypothesis proposed in the introduction. You should authenticate the results obtained by ensuring that they comply with the questions asked or the hypotheses formulated. Each outcome should be discussed, even the one that you did not expect; otherwise, it is useless to demonstrate the hypothesis.

The discussion usually begins with a brief overview of the main findings of your study. You must justify the methods used by comparing them to other methods that you could have used. Although you inevitably have to make methodological choices based on the aim of your study, the time, and the resources available, you need to show the adequacy of your approach for answering your research question. Then you can proceed to discuss the nature of the relationships between the different variables. It is better to follow the same sequence observed in the Results section to interpret each data in relation to the main research question. By going back on the hypothesis, verifying the validity of the results, referring to the authors who approached the studied question, and by showing the relevance of your work, you will be able to make inferences, draw conclusions, develop a theory, or make recommendations which allow science to advance.

In the Discussion section, you should answer the following questions:

– Do your results respond to the hypothesis being tested?
– How do you justify your results?
– Do your results agree or contradict the results of other researchers?
– If so, what is your interpretation?
– Could this conflict be due to the incorrect design of your project or those of other researchers?
– What new and exciting aspect does your research bring compared to what is known?
– What limit did you face?

The researcher must determine how to introduce the previous studies and decide how to proceed with his/her discussion and reasoning. It involves evaluating the entire research process and demonstrating the validity of the results and their relevance to the research problem and hypotheses, relating the results to other studies, and appreciating the limitations of the generalization of the results. In short, the researcher discusses the results of his/her research in the light of previous works and the methods used. Because research deals with discrepancies or inconsistencies, you must deal with conflicting opinions in the "Discussion." Therefore, any contradictory studies should be welcomed, examined, and explained. Lack of bias, linking the results with the topics discussed in the background, providing strategies and suggestions for future research, and determining the role of the results in the advancement of basic and applied sciences are among the things that the researcher must pay attention to in the discussion. Moreover, your discussion must open perspectives and avenues of reflection that allow the subject to be extended. The discussion is, in fact, an inverted pyramid. You start with the details of your findings and will finish with the generalities. Pay attention to not repeating the results, which can weaken your Discussion.

In the Discussion section, it is better to use active pronouns instead of absent pronouns ("we did" instead of "It was done"). Use past tenses when referring to study details, results, analysis, and research background. For example: "Implants in the control group *had* lesser, however, significant Implant probing depth (IPD) reduction at 6 months; this phenomenon might be attributed to the repeated removal of supra-gingival plaque (every fortnight) performed in these sites." When it comes to the meaning of the results, their explanation, or their suggestion, you can start writing in the present tense: "Another possible explanation for this phenomenon *is* a strong Hawthorne effect associated with the bi-weekly visit to the dental office, coupled with the mechanical treatment" [15].

The checklist for your Discussion section should consider the following details:

• The justification of methodology (the reliability and validity of the measuring instruments) and the degree of validity of results.

- Avoiding repetition of the results.
- The interpretation of the main results.
- The project limitations: the experimental conditions, the small sample size, the short duration of the test, confounding variables, any difficulties encountered during the collection, or analysis of data should be reported; otherwise, the reviewers may raise them. You must explain how the research plan or the instruments could have been modified to obtain more reliable results.
- Comparison of the results with previously published studies in the field in terms of similarities and discrepancies.
- The significance of the results and their generalization.
- Questions that arise by this research, recommendations, and advice for future research.
- Clinical application and the practical usefulness of the conducted research.

In the last step, the researcher must reach a reasonable and logical conclusion. In the Conclusion section, an attempt will be made to reflect the general views of the research by relying on the main research question and the issues raised. The conclusion is directly linked to the introduction and the research problem. It gives the reader the satisfaction of having toured the hypothesis posed at the end of the introduction and having answered it. That's why it is often advisable to write the introduction and conclusion at the same time.

For writing a reasonable conclusion:

- Reread the introduction so that the conclusion answers the questions/hypothesis in the introduction.
- Make a synthesis: summarize the main issues dealt with in the development of research.
- Highlight answers to the hypotheses.
- Take care in writing the last sentence because it will leave the final impression on the reviewer/reader.
- Useful expressions to use in composing the conclusion: "In conclusion, we can state that...," "In summary, this work focused on...," and "Overall, it would be appropriate to...."
- Avoid speculation, controversy, or expressions that create doubt: "It would be possible to..." and "It could be suggested...."
- Do not include any bibliographic reference.

Ideally, a conclusion should:

- Answer the research question by considering the main result (never mention things not addressed in work).
- Highlight what you have achieved (specify the consequences of the results of your research).
- Propose new research hypotheses for future studies.

5.8   Acknowledgment/Conflict of Interest

Sometimes you need to declare the conflicts of interest, each author's role in the research process, and funding sources after the Discussion section. This part is also an opportunity to thank individuals/colleagues who cannot be named as the manuscript's author but had some contribution in its preparation.

Authors are required to declare their conflicts of interest. The main conflicts of interest can be identified by following the recommendations of the International Committee of Medical Journal Editors (ICMJE). Generally, any activity or relationship that may influence or give the impression of influencing the study could be considered as a conflict of interest.

5.9   References

The reference list is the last section of a scientific article. A research paper must have at least 25–30 references. In the review articles, the number of references could be more (50–100).

Be sure to use a reference software. If the article you submitted to a journal is rejected and needs to be submitted to another journal, the reference style in the second journal may be completely different from the first one, and the software will help you with a single click. In addition, if you decide to add a new reference to the manuscript after preparing the first draft, the software will automatically update all your references. The principles of reference writing and how to use reference software are given in Chap. 6 of this book.

All references cited throughout the body of the manuscript should be listed following acknowledgment. From the wide range of sources, you should only use those relevant to your research subject. For this purpose, you may select studies that:

- Create a context for your research work that reflects existing knowledge.
- Give an example of your chosen method and justify it.
- Compare your findings, confirm your claims, and interpret the results.

You can consider the following items as a checklist for Reference section:

- There is no misinterpretation of the results of other articles.
- The citation format is based on the university/journal instruction.
- All references that appeared as citations within the manuscript are also present in the reference list and vice versa.
- Articles "in preparation" or "submitted," whose quality is not proven, are not in the bibliography.
- If the number of references is limited in the selected journal, reduce them to the most relevant and recent references.

5.10 Submission of a Manuscript

When you want to submit a manuscript, you must select an appropriate journal covering the subject you are addressing. To do this, you need to see which journal has published more articles related to your research subject. By reviewing the scope of the journals, you can find out if they are more generalist or specialized (target readers). You can find out related journals from the references of your own manuscript. For this purpose, it is better to see in which journals the articles you referred to in your manuscript have been published. The journal you cite most in your references could be appropriate for the submission of your manuscript. Reference managers (EndNote, Mendeley) can help you in this regard. By sorting references in the library by journal in alphabetical order, you can see the journals repeated several times. Specific journals also introduce topics or special editions as their priorities. Remember that journals like *Journal of Oral Microbiology* or *Caries Research* only accept specialized articles in accordance with a definite research area such as oral microbiology or caries, while interdisciplinary journals are more open. One other way to find out the appropriate journal is to use the following links, which give you some journal suggestions based on the abstract of your manuscript:

- https://www.edanz.com/journal-selector
- https://jane.biosemantics.org/index.php
- https://journalfinder.elsevier.com/
- https://journalsuggester.springer.com/

For finding journals that publish review articles, you can search for this type of article in bibliographic databases such as PubMed, Scielo, or Scopus. You can sort article types via advanced search or the Limit tab and see the journals publishing these articles.

If your research significantly impacted the field, you might want to publish your manuscript in a high-impact or top-tier journal with greater visibility. The impact factor (IF) is the average number of citations to the articles published in the journal in the last 2 years. Every year in June, journals covered by ISI listed on the Web of Science are evaluated in terms of IF. The institute publishes the impact factor of journals, which is valid for 1 year. This index is the most important and, at the same time, the most practical index of journal evaluation. The *Journal of Dental Research* (JDR) 2-year impact factor is 6.116, and *Community Dental Health* holds an impact factor of 0.679. You can also use the "SCImago Journal and Country Rank" (SJR) indicator to know the visibility of journals indexed on the Scopus platform. If you are sure about the quality of your work and have enough time, you may want to start sending your manuscript to the highest impact journal and submitting it to a lower-impact journal in case of rejection. Another way to make sure you have a chance to publish your work in a high-impact journal is to send your abstract to the editor of the journal in question. This way, you make a pre-submission inquiry before spending a lot of time organizing your manuscript based on the journal format.

Even if your study is not very impactful, it is better to publish it in a journal indexed in recognized databases in your field, such as Web of Science or PubMed/MEDLINE, or at least be listed in the Directory of Open Access Journal (DOAJ). You may check to see if the journal is published by a publisher with a well-established reputation (Wiley, Elsevier, Springer Nature, Taylor & Francis) or recognized society (IEEE, Institute of Electrical and Electronic Engineers) and how much the publication fees are if it is an open-access journal.

Article printing speed is essential. You should consider the acceptance rate of submitted articles, review speed, online publication time, and publication rate. Some journals are published monthly, while others are published bimonthly or quarterly. *Journal of Dental Sciences (JDS)* is a quarterly journal, *Journal of Dentistry* is a bimonthly peer-reviewed journal, and *Journal of Dental Education (JDE)* is a monthly journal. The publication speed in a high-impact journal such as *Dental Materials* could be much longer than a journal with a lower IF. Open-access journals that require a publication fee may have a faster publication process.

Before submitting your manuscript to a journal, download the "Instruction/Guide for Authors" of that journal and prepare your article in the format required by that journal (e.g., the number of words, the font, article structure, requirement for artworks such as format and resolution, reference formatting, etc.). Some journals impose a limited number of words. Original research reports submitted to the JDR must not exceed 3200 words and 40 references. It is essential to follow these instructions carefully when writing your manuscript. Any deviation can lead to the rejection of the manuscript. Before submitting a manuscript, look at some articles already published in the journal to get an idea of the journal's style.

Today, most journals receive articles electronically. A *cover letter* along with the manuscript will be sent to the journal's editor. Search for the name of the editors and address them directly in your letter. The manuscript's message (significant result, novelty, main implication) will be summed up in the cover letter. Moreover, you will mention the legal notices guaranteeing the content originality, the authors' agreement for publication, and the fact that the manuscript has not been sent to any other journal. Remember that you cannot send your manuscript to two or more journals simultaneously. You may consider providing a list of reviewers in your cover letter in case the journal requests it. Some journals have a copyright/journal publishing agreement form that the corresponding author must fill out and submit.

Once you submit your manuscript, its scientific content should be verified by peer review. If there is no discrepancy with instructions, the manuscript will be assigned to the editor, who will decide to send it for peer review or reject it instantly. Keep in mind that in this stage of manuscript screening by the editor, the editor pays attention to three basic questions: Is the manuscript appropriate for the journal or interesting for the journal's readers? Is the main research question important/novel? Is the study scientifically valid? Usually, at this stage, they examine the title's clarity, design, conclusions, and adherence to the principles of writing recommended by the journal. If you receive a decision letter of rejection at the screening stage (right after the submission), it will typically contain comments from the editor or associate editor. If the manuscript passes the screening stage, it will be sent to two or three expert reviewers in the research area. The review could be single-blind (only review-ers have the names of the authors), double-blind (neither reviewers nor the authors are aware of each other's names), or open (without anonymity).

Reviewers will consider the originality and relevance of your research in the field concerned, its coherence and structure, ethical aspects, and all requirements needed for each section of the manuscript (Abstract, Introduction, Materials and Methods, Results, Discussion, and Conclusion). In the reviewer report, the manuscript's con-tent will be summarized in a short paragraph, and the originality and strengths will be pointed out. Then, the questions, the changes that must be made in each section, and some suggestions will be recommended. Inappropriate methodology or analy-sis, lack of originality, little relevance, not answering the question being researched, and poor drafting are the common reasons for rejecting manuscripts. Also, gram-matical errors that make the sentences incomprehensible and cause the text to be difficult to follow for the reviewer are essential factors. Based on the reviewers' comments, the editor will decide if your manuscript could be accepted without any revision, needs major/minor revision, or should be rejected. You will be informed about the journal's decision by email letter that includes comments from at least two reviewers and the editor.

After receiving the journal decision, in case of major/minor revision, you will be required to respond point by point to reviewers' questions, address their comments, and make clear the modifications made in the manuscript by highlighting them in the text. Do not be irritated by the reviewers' comments; if a sentence is obscure for a reviewer, it may be so to the future reader. The reviewers' comments would increase the quality of your work. Accordingly, you should acknowledge that in your letter of response to reviewers. Answer politely and respectfully whenever you disagree with the reviewer's comment. You will have a deadline for resubmission of the revised version. If you need more time, you should let the Journal know imme-diately, rather than waiting until after the deadline.

If the manuscript is accepted, the corresponding author will receive the galley proof for final minor modifications before providing a digital object identifier (DOI) number, which is like a fingerprint for your manuscript. Figure 5.3 shows different steps from the submission of a manuscript until the final decision.

If your manuscript is rejected, you can submit your work to a new journal by modifying the necessary parts. If your manuscript does not fit the framework of a journal, it does not mean that it will not find its place in another publication. If your manuscript is rejected after peer review, do not ignore the reviewers' comments, and use them to increase the quality of your manuscript before submitting it to another

Fig. 5.3 Different steps from the submission of a manuscript until a final decision

journal. Remember, even if you submit your manuscript to another journal, it may end up in the hands of the same reviewers.

The checklist for manuscript submission should consider the following details:

- Finding appropriate journal
- Preparing the manuscript based on the "Instruction to authors" of the selected journal
- Numbering pages, using subheadings for dividing sections
- Writing short phrases, avoiding jargon words
- Paying attention to spelling/grammar
- Preparing a cover letter
- Preparing a response to reviewers' comments after receiving major/minor revision decisions by addressing each question separately and politely

References

1. R.R. Morães, J.W. Garcia, N.D. Wilson, S.H. Lewis, M.D. Barros, B. Yang, et al., Improved dental adhesive formulations based on reactive Nanogel additives. J. Dent. Res. **91**(2), 179–184 (2012)
2. N.Z. Ratzon, T. Yaros, A. Mizlik, T. Kanner, Musculoskeletal symptoms among dentists in relation to work posture. Work **15**(3), 153–158 (2000)
3. N.L. Nicholas, P. Roshanali, F. Kathryn, Work-related musculoskeletal disorders in dental students: A cross-sectional, pilot study from a UK University Teaching Hospital. J. Musculoskelet. Disord. Treat. **6**(3) (2020)
4. J.S. Ribeiro, E.A. Münchow, E.A. Ferreira Bordini, W.L. de Oliveira da Rosa, M.C. Bottino, Antimicrobial therapeutics in regenerative endodontics: A scoping review. J. Endod. **46**(9), S115–S127 (2020)
5. J. Koyama, K. Fukazawa, K. Ishihara, Y. Mori, In situ surface modification on dental composite resin using 2-methacryloyloxyethyl phosphorylcholine polymer for controlling plaque formation. Mater. Sci. Eng. C **104**, 109916 (2019)
6. F. Tabatabaei, M. Rasoulianboroujeni, A. Yadegari, S. Tajik, K. Moharamzadeh, L. Tayebi, Osteo-mucosal engineered construct: In situ adhesion of hard-soft tissues. Mater. Sci. Eng. C **128**, 112255 (2021)
7. M. Moshaverinia, A. Navas, N. Jahedmanesh, K.C. Shah, A. Moshaverinia, S. Ansari, Comparative evaluation of the physical properties of a reinforced glass ionomer dental restorative material. J. Prosthet. Dent. **122**(2), 154–159 (2019)
8. I. Vázquez-Rodríguez, M. Rodríguez-López, A. Blanco-Hortas, U.A. Santana-Mora, Online audiovisual resources for learning the disinfection protocol for dental impressions: A critical analysis. J. Prosthet. Dent. **124**(5), 559–564 (2020)
9. L. Wang, Z. Gao, Y. Su, Q. Liu, Y. Ge, Z. Shan, Osseointegration of a novel dental implant in canine. Sci. Rep. **11**(1), 4317 (2021)
10. J. Jayaraman, Guidelines for reporting randomized controlled trials in paediatric dentistry based on the CONSORT statement. Int. J. Paediatr. Dent. **31**(S1), 38–55 (2020)

11. M. Durkan, R. Chauhan, N. Pandis, M.T. Cobourne, J. Seehra, Adequate reporting of dental diagnostic accuracy studies is lacking: An assessment of reporting in relation to the standards for reporting of diagnostic accuracy studies statement. J. Evid. Based Dent. Pract. **19**(3), 283–294 (2019)

12. D.-W. Lee, I.-S. Shin, Critical quality evaluation of network meta-analyses in dental care. J. Dent. **75**, 7–11 (2018)

13. M. Garrib, J. Camilleri, Retreatment efficacy of hydraulic calcium silicate sealers used in single cone obturation. J. Dent. **98**, 103370 (2020)

14. M.S. de Melo Soares, C. D'Almeida Borges, I.M. de Mendonça, F.G. Frantz, L.C. de Figueiredo, S.L.S. de Souza, et al., Antimicrobial photodynamic therapy as adjunct to non-surgical periodontal treatment in smokers: A randomized clinical trial. Clin. Oral Investig. **23**(8), 3173–3182 (2019)

15. E.E. Machtei, G. Romanos, P. Kang, S. Travan, S. Schmidt, E. Papathanasiou, et al., Repeated delivery of chlorhexidine chips for the treatment of peri-implantitis: A multicenter, randomized, comparative clinical trial. J. Periodontol. **92**(1), 11–20 (2021)

Chapter 6
Reference Management in Scientific Writing

6.1   When and How Should You Cite a Reference in Your Text?

Research is not just about collecting other researchers' works; instead, the investigator's personal findings constitute a significant part of the research. From a legal and moral point of view, another idea must be referred to its owner, unless such a drastic change occurs in the first one that it is legally and morally incorrect to refer to it, and the first author did not have such a purpose. Failure to do so is considered plagiarism. Plagiarism appeals to the notion of intellectual and scientific honesty. We will talk about plagiarism in more detail within the next chapter.

The researcher's essential role is to analyze and criticize the science. If the researcher quotes the exact phrases of other authors without any change, he/she needs to use quotation marks. Usually, scientific and literary expressions must be quoted directly, or in some cases, we may have to critique and analyze them after quoting the author's phrases. Without quotes, it is plagiarism, even with a bibliographic reference. The best way for avoiding quotation marks is to paraphrase the original text (without forgetting the bibliographic reference). If the researcher does not repeat the exact phrase, he/she does not need to include the phrase in the quotation mark.

Pay attention to the citation of a review article in your manuscript. If a review article provides a relevant overview and its analysis advances scientific knowledge, you can cite it in your manuscript. However, if a review article talks about the results of an article that interest you, read the original research and refer to them. Citing a review article instead of the relevant research articles is called secondhand or secondary referencing, which is allowed but may convey wrong interpretations of results or concepts.

You should refer to other researchers' ideas within the body of the text of your manuscript (bibliographic citation) and at the end (bibliographic reference). You will insert bibliographic citations within the text when you borrow arguments from

© Springer Nature Switzerland AG 2022
F. Tabatabaei, L. Tayebi, *Research Methods in Dentistry*,
https://doi.org/10.1007/978-3-030-98028-3_6

a document or quote a phrase. The bibliographic citations refer to the bibliographic reference placed at the end. All documents in the bibliographic reference must be cited at least once in the text. It is imperative to refer correctly to other articles because, in this way, the reader can easily find the referred article, and the author of the referred article can understand how many references have been made to his/her paper. Moreover, correct reference to an article in a scientific journal would increase the journal's credibility and impact factor.

Each journal has a different way of arranging references, but these methods can be divided into two main categories, *annotated* and *numbered*. It is important to note that there must be a one-to-one correspondence between the way of referring and the way of arranging the references (or the bibliography) in an article. If the numbered method is used for referring, the same method should be used in the reference list. In this chapter, we will explain the reference adjustment (bibliographic citation and the bibliographic reference) based on the (1) annotated method, which includes the author's name and date of publication in the text (reference list in alphabetical order), and (2) numbered method, which is accompanied by inserting a number in the text (the same number will be used at the end in the reference list). In either case, the bibliographic reference is entirely inserted at the end of the text. You may see a mixture of these two methods named *alphanumeric*, which uses a number in the body of the article and numbers in alphabetical order at the end.

When preparing a manuscript or a dissertation, the researcher should be aware of the expected bibliographic style of the journal or university and follow the required style. For example, a university or journal may require the Vancouver style, while another journal demands another style (American Psychological Association (APA), Harvard, etc.). You can find the indications for writing the bibliography and the bibliographic citations in a journal's "instructions to authors." Unfortunately, there is not just one or two bibliographic styles. If you have a limited number of references, it is not difficult to manage them manually. However, in most cases, such as preparing a dissertation or a review article, you will encounter many references. The comments of the supervisor or other authors can make it challenging to handle references manually. In case of rejection, a manuscript must be submitted to another journal with a different reference style. Manually changing the references would not be easy each time you want to submit your manuscript to a new journal or make the required modification in the revised version of the manuscript. Even if the style of the new journal is not different from the previous one, or the journal does not require strict reference formatting (references can be in any style), you need to keep a consistent style to present a coherent document. Familiarity with reference management software would help you in these situations. Therefore, we decided to briefly review the use of two reference management software such as EndNote and Mendeley, which has made it very easy to refer and cite references in the manuscript or proposal. These products have the reference template of most journals. By integrating these tools into the word processors (citation plugins), you can easily extract the bibliography, select the appropriate template, and automatically format citations and bibliographies to the chosen style. You can also pick the style required for each journal, and the software automatically adjusts the reference based on the selected style so that no manual adjustment is needed. These tools may also offer other functions like linking these references with documents saved on the computer.

6.2 The Digital Style (Numbered Method)

Prestigious journals such as *Nature* (*British Dental Journal*) maintain the use of a digital style. The so-called *numbered* or "Vancouver" style includes numerical citations in the text, within the parentheses () or square brackets [], in the middle of a sentence or at the end. The reference numbers may be required to be inserted as superscripts, after/before punctuation.

The criterion for the references' precedence in this method is the order of the numbers in the text. If a reference to an article or a book is cited in a particular section of a manuscript/proposal with a specific number, then the same number must be used in the bibliographic reference at the end of the document (in order of appearance in the text). If multiple sources are referenced in a section of the text, a dash or comma should be used to separate them.

> **Example [1]:**
> "Some studies demonstrated that pressed lithium disilicate veneers showed better marginal adaptation, thinner cement layers, and greater resistance to marginal leakage than those manufactured by a milling process."[8,9]
> 8. Reich S, Wichmann M, Nkenke E, & Proeschel P (2005) Clinical fit of all-ceramic three-unit fixed partial dentures, generated with three different CAD/CAM systems European Journal of Oral Sciences 113(2) 174–179.
> 9. Aboushelib MN, Elmahy WA, & Ghazy MH (2012) Internal adaptation, marginal accuracy and microleakage of a pressable versus a machinable ceramic laminate veneers Journal of Dentistry 40(8) 670–677.

6.3 Author-Date Style (Annotated Method)

In some journals (e.g., *Journal of Dental Research*), the tendency is to the "author-date" style citation principle. One of the most essential and authoritative methods of internal referral is the American Psychological Association (APA) method. This method of citation of the American Psychological Association is related to the rules and regulations established by the American Psychological Association, which refers to the sources used in research articles. In author-date or "Harvard" style, the same pair (author + date) is used for in-text citation and references in the bibliographic list (in alphabetical and chronological order). There must be a complete citation in the list of references for each internal citation and vice versa.

For in-text citation, the names of the authors and the date of publication are stated in parentheses: (First author, 2001). For documents with two authors, both names are indicated (Brown & Dupont, 2010). As soon as there are more than two authors, the first author's name is followed by "et al." (Brown et al., 2013). Two references at the end of one phrase will be separated by ";" (ex. Brown, 2012; Dupont, 2018). If you refer to several documents of the same author published in different years, they should be classified in chronological order, from the oldest to the most recent. But if the articles are published in the same year, the two citations must be explicitly differentiated. The ideal is to add a letter after the year, for example: "(Author, 2012a; Author, 2012b)." These letters added in the citation should be reproduced in the bibliography.

Bibliographic references in this method (the most famous form seen in the APA method) are listed in alphabetical order of author names and in descending chronological order of the publication year. For identical authors and identical years, classify the references in ascending alphabetical order of the words in the title. A single author's reference always precedes the same author's reference(s) when accompanied by one or more co-authors.

Example [2]:
"Qualitative changes also occur due to competitiveness among species (Diaz 2012; Griffen et al. 2012), leading to an increase in pathogenic bacterial taxa. Thus, the dynamic balance among various bacterial taxa is likely to be instrumental in determining periodontal disease activity. Socransky and colleagues (1998) identified combinations of Bacteroides forsythus, Porphyromonas gingivalis, and Treponema denticola as highly associated with clinical measures of periodontal disease."

Diaz, PI. 2012. Microbial diversity and interactions in subgingival biofilm communities. Front Oral Biol. 15:17–40.

Griffen, AL, Beall, CJ, Campbell, JH, Firestone, ND, Kumar, PS, Yang, ZK, Podar, M, Leys, EJ 2012. Distinct and complex bacterial profiles in human periodontitis and health revealed by 16S pyrosequencing. ISME J. 6(6):1176–1185.

Socransky, SS, Haffajee, AD, Cugini, MA, Smith, C, Kent, RL. 1998. Microbial complexes in subgingival plaque. J Clin Periodontol. 25(2):134–144.

6.4   EndNote Software

EndNote is a commercial citation management software that assists you in document collection and organization, article formatting, bibliography creation, and article citations in a wide range of citation methods (Chicago, APA, etc.). This software provides good technical support on its website (http://endnote.com/support/ensupport.asp).

You must have EndNote software installed on your computer to use this software. After installation, if you open a Word document, a section called EndNote should be seen at the top.

After installing the program, by clicking on the EndNote icon, a window will be opened in which you should select the "Create a new EndNote Library" option. From the File menu, click on the New Library icon. Select a location on your computer to save the New EndNote Library and give a name to your Library (e.g., My Thesis). After clicking on Save, your empty library will be ready for the insertion of your references (Note: The "enl" extension is for EndNote files). There is no limit to the number of libraries you can create nor the number of references in each library.

The next step is to enter your references in this virtual library. Each reference has the required bibliographic information such as author name, title, date, and other items. Completing the library in any research is done by using references that have been retrieved while reviewing the literature. Reference entering can be done manually or directly (direct import from databases). It is also possible to transfer stored PDFs to the software to extract the reference's information from the file.

- To enter references manually, click the New Reference icon (or Ctrl + N). A window will be opened, and the Reference Type is "the Journal Article" by default. If you want to add a book reference to the library, first you should change the reference type and then enter the information about it. In this section, through the Import option in the File menu, it is also possible to attach files and images. If you want to edit a reference, you can click on it and make the desired corrections.
- The second way to enter information in an EndNote library is to download information from databases. To transfer articles from Google Scholar (scholar.google.com), you must first select EndNote in Scholar Settings (Bibliography manager). After this step, search for the articles you want. Below each article, there are several links, one of which is "Import into EndNote." If you click on it, a box will appear, and by selecting the "open" option, the article will be transferred to the EndNote file. In the Springer database (https://link.springer.com/), by clicking on the "Cite this article" and then "Download citation," the article will be saved in the format of Research Information Systems (RIS). In the Elsevier database (www.sciencedirect.com), by clicking on the "Cite" link, you can select "Export citation to RIS." At the Wiley site (http://onlinelibrary.wiley.com/), by clicking on the "Tools" link, you may choose the "export citation," and on the next page, you will have access to "EndNote" format. On the PubMed site, after searching, select the articles you want and then click on the "send to" option. In the box that opens, after choosing the Citation manager, click on Create File. A window will open to transfer information to your virtual library by selecting the Open with (EndNote) option and clicking Ok.
- You can also search directly through the EndNote software and download the desired articles from the results obtained. To do this, select the *connect* item from the *file* menu. After picking the desired database, click the *connect* button. Enter

your keyword and start searching by specifying any field, author, title, etc. After entering the keywords, start the search by clicking on the *Search* button. At the end of the search, a window will open showing the number of retrieved articles. By clicking on the *ok* button, the search results will be visible. To save the recovered information, open the *Copy All References To* drop-down option and select the "*choose library*." Select the desired library in the window that opens, and the recycled references will be transferred to the library. Just be sure to use the temporary library method first so that the results are not automatically stored permanently in your library.

– Another way is to drop your stored PDFs in the EndNote library. Their information will be automatically retrieved by software.

Tip 1: In different versions of EndNote software, icons with other names may be used. For example, instead of the "connect" icon in some versions, you may have to click on the "Online Search" icon and select the desired database from the opened page.

Tip 2: Sometimes, you may have duplicate articles in your library on different sites; it is better to remove these duplicates from the EndNote file before you start referencing. To do this, in the reference's menu, click on the "find duplicates" icon; this will find duplicate articles, and you can delete them.

Tip 3: No matter how you save the information in your library, it is possible to edit the information and attach the article file by clicking on any reference. In the attached PDF file, it is possible to highlight some parts and make notes on them.

Tip 4: It is also possible to group saved articles. This feature will be handy when you want to create a review article with different components. You can create specific groups for each section through the Groups menu and by clicking on Create Group.

Tip 5: If you have saved resources for an article in a virtual library and now you want to use some of these resources in another article that has its own library, you can select the resources you need in the first library and Click on the "copy references to" and choose library options in the References menu to move them to the second library.

Tip 6: Once you have researched the articles related to your research and saved them in a library, it is time to choose the right journal to submit the manuscript. Articles stored in the EndNote file are usually arranged in alphabetical order by the authors' names. However, you can click on the word "Journals" on the same main page of the EndNote file. This way, the articles will be sorted by journal name, and you can find out in which journal the studies related to your research are generally published.

When preparing your manuscript in a Word document, you must first determine the standard format of references for your article or dissertation before entering a citation (e.g., APA, Vancouver, etc.). To do this, select "Output Styles" in the EndNote menu, then click on "Open Style Manager." If you think the format of the journal—in which you want to publish the article—is available on EndNote, select the "Select Another Style" item to see a list of available formats. The annotated and numbered methods described earlier are also in this list. Scroll through the window, and after viewing the desired format, select it and click the choose button. If you do not see the format you want, you can download it through the EndNote website or sometimes through the journal's site itself. Otherwise, in the EndNote file, select the "output" item from the Edit menu and create the journal's format or the format recommended by the university for writing the dissertation via "New Style." Additionally, keep in mind that if a journal has modified its previous format, you can select the "edit the currently selected style" icon in "Output Styles" and adjust the format online.

Citing the references in your manuscript can be done using this software in several ways:

– In the Word file, click where you want to refer; click on Insert Citation, then on Find Citation, and browse the article by year, author name, or article title. After selecting the desired reference, click on the Insert button.
– The second method is to select the articles in the EndNote library first, and by clicking on Insert Citation in the EndNote file, the reference will be entered in the Word file.
– You can also enter the record number of the article in the endnotes in parentheses with the # sign at the same time as typing the article (example: {46 #}). EndNote will automatically identify the reference and enter the reference number (in Vancouver format), or the author's name and year (in APA format) based on the selected format.

Utilizing any of the above methods based on the selected citation format will display the citation in the text. EndNote will also automatically add references in the chosen format to the end of your article or dissertation.

Deleting a reference entered in the article must be done through the software. To delete a reference in the text of the article, click next to the reference, and then click on "Edit Citation." Select the desired reference in the window that opens, click on the Remove icons, and then ok.

EndNote sets references in the article using codes called *field codes*. These codes must be removed before submitting the article to a journal or printing the dissertation. To do this, click on the "convert citations icon" and then "convert to plain text" via the EndNote tool at the top of the Word file of the article or dissertation. A warning command tells you that this will delete all the codes. By clicking "ok," a new file with no code will be opened, and you can save it with a new name. Save the previous file containing the codes so that you can apply the corrections considered by the reviewers (which sometimes leads to changes in the references). Correspondingly, if a journal rejects your manuscript, and you want to submit it to another one, by changing the citation format in the file with codes, with one click, all references will be set to the style of the second journal, and the manuscript will be ready for a new submission.

6.5 Mendeley Software

Mendeley, named after the great chemist Mendeleev, is a free Reference Manager and network software that helps you organize research, collaborate with other people on the network, and find the latest scientific research. You can have access to your library by using desktop and web versions. In the web version, you can manage the library online, while in the desktop version, you can manage, bookmark, and cite articles. Mendeley software allows you to attach PDF files to any of the articles entered in the database. This way, you can view the files in the software and take notes on them. You can also share these notes with other people who use this software and are in the same group as you are. It also gives you the ability to search for your keywords in the text of all articles.

In the first step to use this software, you must visit their website (www.mendeley.com), create a free account on this page to register, then download the software and install it on your computer. By signing on to the website, you will have a personal profile. All the information you enter on the desktop will also be accessible online. This way, if the software is accidentally deleted from your computer or you want to access your library information from another computer, you can access it by entering your username and password on the Mendeley website and making corrections to your online library.

After installing the software and opening it, enter your username and password. In the next step, you need to click on the "Install Web Importer" from the "Tools" menu; this will reopen the software site to install Web Importer; click on "Get Web Importer" to add the Mendeley Web Importer icon in the toolbar. This plugin can be used for Google Chrome, Internet Explorer, and Firefox. You must also install the *Mendeley Cite for Microsoft Word* through the software *Tools* menu. After installing it, this software can be seen in the "References" section of your Word file.

A new folder can be opened and named for each article or dissertation to create a virtual library in this software. For example, if you want to start preparing a manuscript about "oral biofilm," by clicking on the "create folder," you can open a new folder and name it as "oral biofilm." You can also define subfolders for each folder to make it easier to categorize your manuscript. In the "my publication" section, you can add your own published articles. Enable the "Sync" option if you want to access your libraries anywhere. Each time you connect to the Mendeley Web and add information, click "Sync" on the desktop to transfer the information.

You can use three methods to import resources into the virtual library:

1. Manually add references: You can enter the DOI or PubMed identifier or PubMed unique identifier (PMID) of the article and click on the magnifying glass icon so that Mendeley can find all the information of the article.
2. Add PDF files of the articles: To do this, you can use the "add document" option of the software itself or select the file and move it to the desired folder in the software (drag and drop). Mendeley can review your files and automatically extract library information from your articles (including article title, author name, journal name, year of publication, etc.). If the software is suspicious of its

data, a message will appear on the right side of the page, which by clicking on search by title, the software will perform a new search to correct the information. You can also make corrections manually in the same menu on the right.

3. Add a folder containing several PDF files: Use the "add folder" option in the desktop version and select a folder containing several articles. You can also use the "Watch folders" option and select a folder to transfer its articles to the library regularly. When you put a new article in the watched folder, it will be automatically transferred to Mendeley.

4. Search for an article on Mendeley's website (papers section). After finding articles, you can "save reference" to your library; see "related research" as well to find new articles.

5. Search for an article on Mendeley desktop (Literature Search): It is possible to connect to dozens of valid search engines and databases through the software to receive reference information directly.

6. Use Web Importer: If you are on a specific site and you want to transfer the articles on this site to the software, click on "Mendeley Web Importer" in the browser toolbar. After a short time, a window will open where you can select the desired articles and then click on "Edit" to specify the folders into which you want the information to be transferred. Click "Save To Library" and share the information with your library. To view it in the software, you might need to click on the "Sync" icon. Each of the scientific sites described for EndNote software can also save information on the Mendeley desktop.

7. Add references that you have already collected by other software such as EndNote: In case you previously saved your articles on other reference management software such as EndNote, you can transfer them to your Mendeley library. To do this, click on "Export" in the "File" section of the EndNote software. Then select the "XML and RIS output" method. Then save your EndNote library in this format. In the second step in Mendeley software, click on the "Add Files" icon. Select the XML files that you previously exported from EndNote and click "Open." EndNote library information will be transmitted to Mendeley.

Tip 1: Before you start referencing, be sure to remove duplicate references from your library. If some of the articles are transferred to the software in several ways and are duplicates, you can select them together (from the Tools menu, click "Check for Duplicates") and then click on the "Merge documents" option. You will be taken to a page where a new file is obtained from a combination of previous files, and in the following rows, there are selected files; then, you can confirm the merge.

Tip 2: The article information in the software has a DOI/PMID code and a search mark in front of it; you can access more information from the article by clicking on the search mark.

Tip 3: You may write comments in each article file, highlight the desired sections, search for the article in the software, and email it through the software (using the share option). Furthermore, each article you download will have a green dot next to it by the software indicating that you have not yet read the PDF file, and this dot will be removed after reading it. If you are interested in an article and want to refer to it later, click on the asterisk next to it. In this way, the article will be automatically transferred to the Favorites folder, and you can read the article later by referring to this folder. Also, if you have another file, such as a Word file or PowerPoint related to an article, in the article information section (files section), you can click on Add file and add other files next to the PDF.

Tip 4: In the search section of the library, by clicking on the magnifying glass icon, you can search for the word you want in the title of the articles, the names of the authors, etc. The literature search section on the left shows the search for other people's Mendeley files. By right-clicking on any article in your file and then clicking on related documents, you can find articles similar to that article in Mendeley.

How to cite articles in your Word file via Mendeley? First, you must select the *citation style* (such as Vancouver, Harvard, etc.) through Mendeley software's "view" menu. Choosing the citation style through the "References" menu in the Word file is also possible. If the required style is not available in your list, you can click on "Get More Styles" in "More Styles" and type the journal's name in the search field and install the style you want. If you need to edit a style, click on "about" and then on the CSL editor and make the desired corrections in the same window that opens.

Click "Insert Citation" wherever you want to add a reference from the "References" menu to cite articles. A window will open, and you can search for references by author's name, year, keywords, etc. Click on "ok" to enter the reference. References will not be added to the end of your text at the same time; instead, at the end of the work, wherever you want to add references, click on "Insert Bibliography" through the "References" menu in the Word file, and all the required information will be entered.

It is also possible to cite references through Mendeley software. To do this, after clicking on "Insert Citation," click on "Go to Mendeley." The Mendeley software will open, and after selecting the article, you must click on the "cite" icon. You can also add multiple resources at the same time by holding down the Ctrl key.

Deleting a reference in this software is easy with the keyboard, and there is no need to delete it through the software. If you enter a new reference between references, the number of all references will be changed automatically.

If you want to remove Mendeley codes from your file, in the "References" section of your Word file, you can click "Export as" and then click "Without Mendeley Fields." The file will be disconnected from Mendeley. Keep a copy of the original file with codes.

It is also possible to add Word files to the library using the "Add file" option. Still, you can prevent the file from being downloaded by others by clicking on the "unpublished work" in the file specifications in the "other setting" section.

With Mendeley, it is possible to create private groups to share documents in addition to the references of these documents. You can share your bibliography with groups and share information with colleagues. To create a group, you may click on the "Create Group" in the software, then invite three people to join (you must pay to increase the number in a group or create more groups). Each group member can upload articles to the group. Your highlights and notes on each article will be shared with other members of the private group. Each member of the group has a different color to highlight. The "Overview" option in the group summarizes what has happened so far (articles added and conversations exchanged between group members). You can leave a comment on this page.

References

1. I. Soares-Rusu, C.A. Villavicencio-Espinoza, N.A. de Oliveira, L. Wang, H.M. Honório, J.H. Rubo, et al., Clinical evaluation of lithium disilicate veneers manufactured by CAD/CAM compared with heat-pressed methods: Randomized controlled clinical trial. Oper. Dent. **46**(1), 4–14 (2021)
2. D.T. Graves, J.D. Corrêa, T.A. Silva, The oral microbiota is modified by systemic diseases. J. Dent. Res. **98**(2), 148–156 (2019)

Additional Resources

https://clarivate.com/webofsciencegroup/support/endnote/
https://www.mendeley.com/guides

Chapter 7
Ethics in Dental Research

7.1 Why Do We Need Ethics in Research?

Dentistry is advancing in various fields such as genetics and gene therapy, tissue engineering, organs-on-a-chip, biomaterials with immunomodulatory properties, smart biomaterials, nanotechnology, and artificial intelligence. New questions are being asked about the moral responsibility of using these techniques in animal and human research with these advances. In addition, many of the ethical challenges already posed have not yet been addressed (such as the time required to follow up with patients who have participated in a study involving a new material or technique or using animals to overcome aging). In addition to these potential problems, ethical consideration in writing research findings is also widespread, and failure to address some issues at this stage can sometimes endanger patients' health. For this purpose, the following moral principles can be taken into consideration:

1. Honesty: Observance of honesty concerning patients who have participated in the research, collected data, and reported results
2. Neutrality: Observance of neutrality and lack of prejudice in justifying patients and interpreting and writing results, disclosing personal and financial interests that may affect research
3. Accuracy: Observance of accuracy in the process of performing research, collecting data, and writing results
4. Criticism: Sharing data and ideas and willingness to accept criticism
5. Respect for the rights of others: The rights of patients, other researchers, copyright, and other individual rights
6. Responsibility: Conducting research in the interests of society, performing well-organized animal research, and not publishing similar articles for personal gain

© Springer Nature Switzerland AG 2022
F. Tabatabaei, L. Tayebi, *Research Methods in Dentistry*,
https://doi.org/10.1007/978-3-030-98028-3_7

7. Observance of justice: Lack of discrimination between patients, fairness towards the article's authors

Indeed, along with the new moral laws and policies established in scientific societies and governments, individual conscience and responsibility can lead the researcher to observe ethics at all stages of research and guarantee research values.

7.2   Ethics in Human Studies

At present, placing absolute confidence in animal models as indicators of physiological, pharmacological, or toxicological effects in humans is impossible. In most cases, the harmlessness effect of a new material or treatment in animal studies does not guarantee that it is safe for humans. Whether diagnostic, prophylactic, or therapeutic, all innovative scientific interventions should ultimately be evaluated in human subjects. ADA's seal of acceptance requires carrying out clinical trials before marketing products in the USA [1].

Human subject means "...a living individual about whom an investigator (whether professional or student) conducting research obtains (1) data through intervention or interaction with the individual, or (2) identifiable private information" [45 CFR 46.102(f)]. Human studies are not limited to clinical trials but also saliva, gingival crevicular fluid, stem cells, tooth, tissue samples, and histological slides.

It is undoubtedly essential to observe ethics in human research; the question is: What rules and regulations are needed to comply with ethical considerations in human studies? Several codes have emerged to ensure the protection of human subjects, including the Nuremberg Code [2], the Declaration of Geneva [3], and the Declaration of Helsinki [4].

In June 1964, at the 18th assembly of the World Medical Association (WMA) in Helsinki, a code of ethics was drafted to guide physicians engaged in research involving human subjects, known as the Helsinki Statement I. At its 29th session in Tokyo in 1975, the same association amended the statement and adopted the Helsinki Statement II. The last adaptation was made in 64th WMA General Assembly, Fortaleza, Brazil, in October 2013. "Ethical considerations regarding health databases and biobanks" were revised by the 67th WMA General Assembly (Taipei, Taiwan, October 2016), and a statement was also adopted by the 71st WMA General Assembly (Cordoba, Spain, October 2020) on "human genome editing" [5]. The above efforts highlight the importance of respecting human rights in research.

In dentistry, the Committee on Ethics in Dental Research of the International Association for Dental Research (IADR) adopted the Code of Ethics in May 2009 [6]. This code of ethics includes human research, animal research, conflict of interest, and dissemination of information. The ADA and FDI World Dental Federation also provided manuals for dental ethics.

The general spirit of the international statements on the observance of ethical standards is respect for the free will of the subject, dignity, rights and welfare, avoidance of possible harm to the subjects or justifying the potential damage by derived benefits, honest and fair treatment of the subject, and commitment of researchers to the confidentiality of the information they receive at the beginning of or during research on the subject. Conducting conscious research while maintaining ethical principles based on Helsinki's statement should be one of the goals of all researchers. The guidelines of the Helsinki Codes now should be respected in all "biomedical research involving human subjects."

Human research usually recruits healthy or volunteer patients. Voluntary participation in the study is the key to the moral debate in human research. Accordingly, the patient should be fully aware of the treatment or intervention and consciously accept this participation. Ethical research involves obtaining the informed consent of those you intend to interview, question, or experiment with. However, many individuals, including children and adults who are mentally ill or intellectually disabled, and those not familiar with modern medical concepts, are incapable of giving informed consent. Therefore, informed consent should always be complemented by an independent ethical review of research proposals. The ethics review boards should check that all the interventions proposed are of an acceptable level of safety to be undertaken on human subjects and ensure that all other ethical considerations raised by the protocol have been satisfactorily resolved in the proposal.

The Institutional Review Board (IRB) should review and approve the research proposal, informed consent forms, and documentation of "Research Compliance Training" and no "Conflicts of Interest" before conducting research. All proposals related to human subjects should include the results of relevant laboratory and animal research, the expected benefits and potential risks to participants, the full details of the informed consent procedure, the evidence of the researcher's qualification and experience, the arrangements for protecting the confidentiality of information, and research's accordance with the Helsinki Declaration. The researchers cannot deviate from the proposal approved by the IRB. Accordingly, the final report or manuscript must mention that the study was conducted according to the Declaration of Helsinki, the local ethical committee approved the proposal, and the patient's informed consent was received.

7.3   Ethics in Animal Studies

We need to establish the relevance of in vitro experiments by in vivo studies. Laboratory animals are special specimens in approved suppliers used for the pathogenesis of oral diseases, the biocompatibility of new dental materials, or the appropriateness of new treatments. An animal (derived from the Latin word animus) is a living being that does not have human characteristics yet is able to feel. The animal is a variable whose changes depend not only on its hereditary characteristics but also on environmental factors. A healthy animal will be selected to perform the

experiment and predict what will happen to the human body based on the anticipated characteristics.

Until the mid-nineteenth century, animals were mainly used to learn the anatomy
and physiology of the body. With the advancement of microbiology, the discovery
of hormones, vitamins, and the pharmaceutical industry, more research has begun
on animals. Millions of animals are currently being used in research laboratories.

Ethics in animal studies is more dependent on the personal moral responsibility
of the researcher. However, individual conscience and responsibility do not seem to
be effective alone. Therefore, legislating regulations can ensure that animal rights
are better respected.

Russell and Burch examined decision-making mechanisms for the use of animals in research. They suggested that any animal research should be evaluated in
terms of compliance with the three principles: replacement, reduction, and refinement [7].

- *Replacement*: Animal experiments should be replaced—as much as possible—
  by developing alternative methods (such as laboratory tests, 3D in vitro models,
  or computer programs). Sensitive animals (such as vertebrates) can be replaced
  with less sensitive animals (non-vertebrates) or tissue-engineered models.
  Current advances in three-dimensional tissue engineering, lab-on-a-chip, and
  organ-on-a-chip can be effective in this aspect.
- *Reduction*: With careful planning before testing, the minimum number of animals required to get valid results can be tested.
- *Refinement*: Good laboratory conditions should be provided for the animal to
  reduce its suffering and increase its well-being (adequate living conditions, minimizing pain on distress, developing anesthesia methods, using non-invasive techniques, and providing postoperative care).

In addition to these three principles, the following points must also be considered:

- Animal experiments should only take place when no alternative is possible.
- The choice of animal species to be tested should be based on the purpose of the
  research.
- Endangered animal species (in the red list) should only be tested with the purpose of conserving the species.
- The suitability of the experimental proposal should be proven (by reviewing
  articles and discussing with colleagues from other disciplines).
- At the end of the experiment, wild animals should be released into the prey if
  they can survive. Otherwise, the death of the animal must be in a completely
  moral way.
- It is essential to write an accurate proposal before starting work. In this proposal,
  objectives and assumptions, animal models, the number of animals required, the
  method of analysis, and how the results are exploited should be clearly stated and
  specified.

- The Institutional Animal Care and Use Committee (IACUC) must review the proposal to verify if ethical guidelines are being followed and oversee animal experiments.

It is estimated that 115 million animals, including rodents, such as mice and rats, as well as dogs, pigs, and monkeys, are used for dental research purposes worldwide each year [8]. A recent study revealed a doubling of the number of publications of animal experiments between 1982–1983 and 2012–2013 in two representative and high-level periodontal journals (the *Journal of Periodontology* and the *Journal of Clinical Periodontology*) [9]. The rational use of animal experiments, the application of alternative methods (3D models), or a combination of the two approaches could reduce the number of animals in experiments.

In September 1989, at the 41st World Medical Assembly (WMA), the statement on animal use in biomedical research was adopted, revised in 2006, and reaffirmed in 2016. The ethical guideline published by the American Psychological Association (APA), which offers advice on the welfare of animals used in research, supports high-quality research by ensuring that researchers follow sound reasoning and that the methods used are scientifically, technically, and morally appropriate.

ISO 10993 (1-20) standards are for studying the biological properties of materials used in medicine. Among these standards, ISO 10993-2 on animal welfare includes animal tests' requirements [10]. The required tests (cytotoxicity, genotoxicity, sensitization, and implantation) could be selected based on the nature of the contact of biomaterials with the body and the contact duration. A new orthodontic bracket that is a surface device in contact with teeth does not need to be evaluated by implanting in an animal's body. Also, an impression material that is in contact with prepared teeth (external communicating devices) does not need genotoxicity and implantation tests due to limited exposure (<24 h). The key provisions of ISO 10993-2 guideline are:

1. Promoting science regarding the higher application of non-animal tests and ensuring that animals are treated ethically
2. The need for compliance with animal care standards, staff experience, and minimizing animal pain
3. The rules for designing the study: include objectives, methods, number of animals, species, source of animal supply, animal health status, reuse of animals, and how the animal dies

Standard animal tests in dentistry are mucous membrane irritation (for evaluating inflammation), skin sensitization (for evaluating allergy), implantation (for evaluating materials in contact with subcutaneous tissue or bone), systemic toxicity, inhalation, dental pulp irritation, mucosal and gingival usage tests, and endodontic usage test. Some tests that evaluate a substance regardless of its final use in the human body can be performed in animals like mice, rats, hamsters, or guinea pigs. However, larger animals like dogs and monkeys are required in the usage tests (dental pulp irritation, pulp capping/pulpotomy, and endodontic usage) that assess a substance concerning its final use and its intended clinical application [11]. The latest standard review of ISO 7405 (Dentistry — Evaluation of biocompatibility of medical devices used in dentistry) suggests that alternative methods should be used for animal testing to assess the biocompatibility of dental materials [12]. However, it should be noted laboratory tests cannot replace all types of animal and human studies. Tests such as laboratory toxicity tests do not consider the complex interactions of the material with the body during implantation. Combining laboratory, animal, and human studies with full consideration of ethical rules while following the ARRIVE (Animal Research: Reporting of In Vivo Experiments) guidelines and the SYRCLE (Systematic Review Centre for Laboratory animal Experimentation) risk of bias tool can be the most effective way to test the new treatments and dental biomaterials.

Paying attention to the importance of respecting animal rights does not mean opposing animal testing. But every researcher should make sure that there is no abuse in this regard. Although animals are often housed in research laboratories for research purposes, this does not mean that we can treat them in any way we want. Every animal is a sensitive creature and should be treated with this in mind. Many people ask themselves: What is the need to take care of animals in our experiments while many people worldwide are suffering from diseases? The answer is straightforward: Those who hurt animals will certainly not hesitate to do the same with human beings. Hence, humans will not be at peace as long as animals and nature are harmed; and it must be acknowledged that respect for nature and animals is linked to our existence.

7.4 Ethics in Manuscript Writing

Research ethics is dependent not only on the way we treat humans or animals but also on how results are reported. The lack of ethics in writing research results can make our research invalid. A nonprofit organization called the Committee on Publication Ethics (COPE) was set up in 1997 to investigate and report on ethical issues of scholarly publishing [13]. It is important to note that in many cases, insufficient awareness leads to non-compliance with moral issues. For example, many students do not know the correct way to refer, leading to plagiarism. Therefore, proper knowledge of "how to write articles" and "what the ethical problems in writing are" can play an essential role in reducing these problems. In this section, we are referring to the most critical ethical issues in reporting research results.

- Authorship

Observance of ethical principles in writing authors' names based on their participation in a project is one of the basic principles for writing a scientific article. Unfortunately, sometimes people who have not participated in answering the research question or made a significant contribution to the research project are mentioned as authors in an article. According to the International Committee of Medical Journal Editors (ICMJE), you must play a role in all three of the following areas to be included in the list of authors. Anybody who does not meet all these conditions should not be included as an author in a manuscript:

1. Presenting ideas, research design, implementation, data collection, data analysis, or data interpretation
2. Drafting the manuscript or revising it critically
3. Approving the final version for publication in the journal and agreeing to be accountable for the full content of the article

The NIH also provides a guideline with a graphical scale to address authorship contribution [14]. Remember that playing a role in only one of the three authorship requirements (such as data analysis), funding research, research group supervision, technical help, or writing assistance is not an excuse to add names to the list of authors. A person who does not conform to all three conditions should not be mentioned as an author but could be mentioned in the *Acknowledgment* section of the manuscript. Nowadays, some journals call for defining the role of authors and their contributions to the manuscript by submitting an "author statement file."

• Fraud (Data Fabrication/Data Falsification/Omission of Data)

Lack of honesty in research includes data fabrication (making up the data), manipulation of results (changing or adding data), or failure to provide all the results obtained. You can neither falsify nor exclude data in favor of your hypothesis. Do not exclude some data, which are inconsistent with others, from the analysis unless you doubt their credibility based on a valid reason (you should report this omission). Referring to the results that are not supporting the desired outcome is an excellent moral practice, while the selective report of data is inadequate. If the data of a scientific document cannot be wholly reproduced under the same conditions, it indicates data fabrication.

• Undeclared Conflict of Interest (COI)

Conflict of interest (COI) includes situations that can affect the judgment of reviewers or the editorial board. Any result or discussion mentioned in the article that benefits one of the authors is part of the conflict of interest. This conflict of interest should be resolved by declaring the person's name in the acknowledgment section. Authors are required to disclose any interests that may affect research and acknowledge funding from universities, organizations, companies, and other sources that the publication of results can benefit or harm. The interests may be personal, commercial, political, or financial. Many organizations are also conducting research studies for internal purposes. Therefore, the findings have exclusive rights, and their publication without acknowledgment of funding sources is immoral.

• Plagiarism

Plagiarism is derived from the Latin word "plagiarius," meaning "kidnapper." According to the Merriam-Webster dictionary, plagiarism is "the act of using another person's words or ideas without giving credit to that person." It is not morally acceptable to present the work of other researchers without referring to them. In the same way, authors should not use the exact phrases of their previous articles (self-plagiarism) unless they refer to them. The direct and indirect quotations and how to cite them were discussed in detail in Chap. 6. Currently, journals have various anti-plagiarism software such as WCopyfind, iThenticate, CrossCheck, or Turnitin to detect plagiarism. The decision of a journal will vary based on the degree of plagiarism.

Copying a figure or a table from a published work is also illegal and is called copyright infringement. Before taking such an action, the authors must contact the publisher or the author concerned and obtain permission (copyright permission).

- Duplicate Submission/Redundant Publication

It is not morally appropriate for an author to simultaneously submit a manuscript to more than one journal or publish a similar article (resemblance in hypothesis, data, discussion, or conclusion). When submitting a manuscript to a journal, the corresponding author declares that they did not submit it to another journal at the same time. You can send the manuscript to another journal after receiving the rejection letter or withdrawing your manuscript from the first one. Presenting a summary of a manuscript during conferences or meetings does not include this ethical problem. When using data from your previous study, even if you are providing additional data in the new manuscript, it is a redundant publication. The redundant publication includes splitting the results of a single project into several parts and preparing several manuscripts based on that.

References

1. A.M. Mark, Smart shopping with the ADA Seal of Acceptance. J. Am. Dent. Assoc. 151(1), 72 (2020)
2. J. Katz, The Nuremberg code and the Nuremberg trial: A reappraisal. JAMA 276(20), 1691 (1996)
3. Declaration of Geneva, J. Nutr. Med. 3(2), 153–153 (1992)
4. B. Gandevia, A. Tovell, Declaration of Helsinki. Med. J. Aust. 2(8), 320–321 (1964)
5. C. Kurihara, V. Baroutsou, S. Becker, J. Brun, B. Franke-Bray, R. Carlesi, et al., Linking the declarations of Helsinki and of Taipei: Critical challenges of future-oriented research ethics. Front. Pharmacol. 11, 1692 (2020)
6. International Association for Dental Research, Code of ethics for dental researchers. J. Am. Coll. Dent. 81(3), 19–22 (2014)
7. J. Tannenbaum, B.T. Bennett, Russell and Burch's 3Rs then and now: The need for clarity in definition and purpose. J. Am. Assoc. Lab. Anim. Sci. 54(2), 120–132 (2015)
8. V. Nagendrababu, P.E. Murray, A. Kishen, M.H. Nekoofar, J.A.P. de Figueiredo, P.M.H. Dummer, Animal testing: A re-evaluation of what it means to Endodontology. Int. Endod. J. 52(9), 1253–1254 (2019)
9. N. Staubli, J.C. Schmidt, C.A. Rinne, S.L. Signer-Buset, F.R. Rodriguez, C. Walter, Animal experiments in periodontal and Peri-implant research: Are there any changes? Dent. J. 7(2), 46 (2019)

10. ISO, *ISO 10993-2:2006 Biological Evaluation of Medical Devices – Part 2: Animal Welfare Requirements* (International Standard, 2006)
11. R. Sakaguchi, J. Powers, *Craig's Restorative Dental Materials* (Elsevier, 2012)
12. ISO. Evaluation of biocompatibility of medical devices used in dentistry – ISO 7405. Am. Natl. Stand. Dent. 2008
13. E. Wager, The Committee on Publication Ethics (COPE): Objectives and achievements 1997–2012. Presse Med. **41**(9), 861–866 (2012)
14. https://oir.nih.gov/sites/default/files/uploads/sourcebook/documents/ethical_conduct/guidelines-authorship_contributions.pdf

Additional Resources

https://www.wma.net/policies-post/wma-declaration-of-helsinki-ethical-principles-for-medical-research-involving-human-subjects/

Index

Printed in the United States
by Baker & Taylor Publisher Services